T0301405

Academic Spin-Offs and Technology Transfer in Europe

Academic Spin-Offs and Technology Transfer in Europe

Best Practices and Breakthrough Models

Edited by

Sven H. De Cleyn

iMinds vzw, Ghent and University of Antwerp, Belgium

Gunter Festel

FESTEL CAPITAL, Fuerigen, University of Basel, Switzerland and Technical University of Berlin, Germany

Cheltenham, UK • Northampton, MA, USA

Published by
Edward Elgar Publishing Limited
The Lypiatts
15 Lansdown Road
Cheltenham
Glos GL50 2JA
UK

Edward Elgar Publishing, Inc.
William Pratt House
9 Dewey Court
Northampton
Massachusetts 01060
USA

A catalogue record for this book
is available from the British Library

Library of Congress Control Number: 2016942154

This book is available electronically in the **Elgar**online
Business subject collection
DOI 10.4337/9781784717384

ISBN 978 1 78471 737 7 (cased)
ISBN 978 1 78471 738 4 (eBook)

Typeset by Servis Filmsetting Ltd, Stockport, Cheshire
Printed and bound in Great Britain by TJ International Ltd, Padstow

Contents

CONCLUSION

Figures

Tables

Contributors

Dr Rajiv V. Basaiawmoit is Head of Sci-tech Innovation & Entrepreneurship at Aarhus University and co-founder of Biosymfonix (rvb@au.dk).

Dr Julia Bauer is responsible for the business ideation phase within the Fraunhofer Venture Lab (julia.bauer@fraunhoferventure.de).

Dr Michael Brandkamp is CEO of the German seed stage investor High-Tech Gründerfonds (m.brandkamp@htgf.de).

Professor Dr Marco Cantamessa is Professor at Politecnico di Torino and Chairman and CEO of I3P, the incubator of Politecnico di Torino (marco.cantamessa@polito.it).

Professor Dr Sven H. De Cleyn is Director Incubation at iMinds and Professor in Entrepreneurship at University of Antwerp (sven.decleyn@iminds.be).

Dr Gunter Festel is founder and CEO of the investment company FESTEL CAPITAL (gunter.festel@festel.com) and engaged with University of Basel and Technische Universität Berlin.

Professor Dr Flemming K. Fink is Director at the AU Centre for Entrepreneurship and Innovation at Aarhus University in Denmark (fkfink@au.dk).

Professor Dr Frank Gielen is Director Professional School at iMinds and EIT Digital, as well as Professor at Ghent University (frank.gielen@iminds.be).

Professor Dr Victor A. Gilsing is Full Professor at Faculty of Applied Economics at University of Antwerp (Belgium) and at Centre for Innovation Research (CIR) at Tilburg University (the Netherlands) (Victor.Gilsing@uantwerpen.be).

Mia L. Justesen is development consultant at University College Nordjylland (mlj@ucn.dk).

Matthias Keckl heads the Fraunhofer Venture Lab team and is responsible for the development of innovation concepts in the fields of intra-/entrepreneurship (matthias.keckl@fraunhoferventure.de).

Dr Florian Kirschenhofer is start-up manager at Max-Planck-Innovation (kirschenhofer@max-planck-innovation.de).

Professor Dr Magnus Klofsten is Professor of Innovation and Entrepreneurship at Linköping University, Sweden (magnus.klofsten@liu.se).

Professor Dr Jan Kratzer heads the Chair of Entrepreneurship and Innovation Management at Technische Universität Berlin and is managing director of the Center for Entrepreneurship at Technische Universität Berlin (jan.kratzer@tu-berlin.de).

Dr Thorsten Lambertus is part of the Fraunhofer Venture Lab team and heads its acceleration program FDays (thorsten.lambertus@fraunhofer-venture.de).

Dr Hervé Lebret is senior scientist and head of Innogrants, a pre-seed fund supporting entrepreneurship at École Polytechnique Fédérale de Lausanne (EPFL) (herve.lebret@epfl.ch).

Dr Erik Lundmark is senior lecturer in the Department of Marketing & Management at Macquarie University, Sydney, Australia (erik.lundmark@mq.edu.au).

Ulrich Mahr is a member of the General Management of Max-Planck-Innovation, the technology transfer office of the Max-Planck-Society (mahr@max-planck-innovation.de).

Kirstine V. Moltzen is Program Manager of the AU Student Incubator (SVAA) at Aarhus University (kvm@au.dk).

Dr Matthias Mrozewski is Assistant Professor at the Chair of Entrepreneurship and Innovation Management at Technische Universität Berlin (matthias.mrozewski@tu-berlin.de).

Professor Dr Isabelle M.M.J. Reymen is Assistant Professor in Innovation Technology Entrepreneurship & Marketing at the Technical University Eindhoven (i.m.m.j.reymen@tue.nl).

Professor Dr A. Georges L. Romme is Full Professor in Innovation Technology Entrepreneurship & Marketing at the Technical University Eindhoven (A.G.L.Romme@tue.nl).

Björn Schmalfuß is a member of the Fraunhofer Venture Lab team responsible for the development of the Fraunhofer Ideenportal as the new innovation management system for Fraunhofer (bjoern.schmalfuss@fraunhoferventure.de).

Dr Helmut Schönenberger is co-founder and CEO of UnternehmerTUM, the center for innovation and business creation at Technische Universität München (schoenenberger@unternehmertum.de).

Professor Dr Elco van Burg is a part-time associate professor of Entrepreneurship and Organization at VU University Amsterdam (j.c.van.burg@vu.nl).

Agnes von Matuschka is managing director of the Center for Entrepreneurship at Technische Universität Berlin (agnes.matuschka@tu-berlin.de).

Acknowledgements

We are sincerely grateful to all those who contributed to this book, particularly our co-authors who carved out time in their busy academic and business lives to share their vast experience of creating and supporting academic spin-offs and spurring technology transfer activities in their region or at their respective institutions. Our heartfelt thanks to each and every one of them. Their dedication to this project has been an inspiration and their insights have proven very valuable. It has been our privilege and pleasure to work with them all.

Special thanks must also be given to Liz Morrison, who spent numerous hours revising drafts of each and every chapter, introduction and conclusions.

We are also grateful for the support received by our respective organizations, iMinds and FESTEL CAPITAL, in realizing this project.

Finally, thank you to Benedict Hill and Francine O'Sullivan at Edward Elgar Publishing for their guidance and assistance in helping us to navigate the course of bringing a collective idea into a published volume.

Sven H. De Cleyn and Gunter Festel

Introduction: what is the current state of knowledge transfer at research institutions in Europe, what are the main challenges and why does it matter?

Sven H. De Cleyn and Gunter Festel

THE KNOWLEDGE TRANSFER SCENE AT PUBLIC RESEARCH INSTITUTIONS

Research institutions, including universities and others, have engaged in knowledge transfer outside their boundaries for centuries, mostly through education and scientific publications. In recent decades, they have tried to find other, more direct ways to bring new knowledge to business and society, often facilitated by a technology transfer office (TTO) or industrial liaison office (ILO). As such, research organizations have been increasingly engaging in more entrepreneurship-related activities: establishing spin-off ventures, setting up investment funds, etc. This additional role has been described as the third mission of research institutions alongside research and education (Etzkowitz, 1998).

However, TTOs most commonly adopt a technology-push or inside-out approach, where new knowledge is 'pushed' from the research institution towards third parties. Common methods applied in this regard include the sale of intellectual property (IP), licensing or creating spin-off ventures to commercialize new technologies. This technology-push approach imposes specific challenges and many research institutions struggle to derive sufficient benefits from their knowledge transfer activities, both in the short as well as the long term.

Deriving sufficient (or more) benefits from knowledge transfer is far from a trivial concern. Several hurdles can hamper the successful transfer of knowledge from research institutions to private companies (or wider still, from one research organization to another). The first important factor impeding knowledge transfer is the cultural difference between the supplying organization and the receiver. The receiver is often a commercial organization (for-profit company) with a very different set of objectives to the supplier. The mindset of both organizational types is so dissimilar it often leads to a difficult handover of knowledge. Secondly, the knowledge being transferred rarely comes as a final product or ready-to-use package. Tacit knowledge typically plays a crucial role in successfully applying new knowledge. However, transferring tacit knowledge is time- and resource-intensive. The third hurdle is the nature of research institutions. These organizations are typically 'engineered' to conduct research (and eventually pass on knowledge through education and scientific publications). However, they are usually not designed for frequent interactions with industry, or for engaging in other forms of knowledge transfer. Equally, a low emphasis is placed on knowledge transfer activities in the academic career path and incentive system. One last hurdle may be the main contributing factor hindering research institutions from successfully engaging in knowledge transfer activities. Many of these organizations have been around for centuries, whilst knowledge transfer activities towards business and society have only emerged in recent decades. Therefore, research institutions typically lack a tradition of and experience in these activities.

Through continuous learning and trial and error, research institutions are seeking new ways of mastering knowledge transfer in order to derive the fullest benefits from it and the process of transferring knowledge and technologies to industrial actors (and broader society) is becoming better understood. Along with increased awareness of cluster thinking that involves academic research, the role of knowledge and technology transfer in the creation of new, technology-based ventures is now recognized. However, current practices and studies still focus on traditional ways of doing this, i.e. from a technology-push orientation (Siegel et al., 2003a) such as licensing IP rights to third parties and/or the creation of spin-off ventures. This approach has proved successful in a number of domains, including biotechnology, pharmaceuticals and microelectronics. However, in other technology domains, especially in those where patents as the main mechanism to protect IP are less common, success is limited (Santoro and Bierly, 2006; Guston, 1999).

The technology-push approach has also proved unsuccessful when it comes to start-ups and small and medium-sized enterprises (SMEs). Both types of company face important challenges in gaining access to the latest knowledge, state-of-the-art technologies and research developed at research institutions (Nunes et al., 2006). SMEs manage knowledge differently from large companies, and research institutions are (often) not adapted to interact with them. While large companies are plugged into the research community or have their own in-house R&D departments, the innovation efforts of start-ups are usually driven by small teams operating without a precisely defined roadmap. In essence, start-ups often lack the 'R' part of the R&D equation (except for start-ups in domains such as biotechnology).

For most (Western) economies, SMEs are seen as the engines of economic growth and innovation. They typically account for at least 50% of employment generated in these economies and represent more than 90% of total economic activity in any region (Federation of Small Businesses (FSB), 2014; International Finance Corporation (IFC), 2012; Singh et al., 2010). Therefore, the difficulty they face in accessing the latest technologies is a real problem.

KNOWLEDGE TRANSFER EMBEDDED IN OTHER PHENOMENA

The European Innovation Paradox

The *Green Paper on Innovation* (European Commission, 1995) first raised the issue of a so-called European (Innovation) Paradox. According to a number of subsequent studies (e.g. Tijssen and van Wijk, 1999; Klofsten and Jones-Evans, 2000; Clarysse et al., 2002), Europe lags behind other world regions (mainly the US) when it comes to transferring research results to the market (successfully translating these results into products and services for commercial and/or societal benefit). This poor record is, among others, attributed to the more competitive US academic environment (mainly in terms of remuneration, promotion and job mobility) and legislative system; a more pronounced distinction between teachers and researchers in Europe; and the relatively predetermined academic career path in Europe that is not conducive to commercialization activities. Other studies have pointed to the strong public sector science base in the European Union (EU) coupled with poor R&D activities in EU firms (Tijssen and van Wijk, 1999). However, a number of other studies have challenged the existence of a European Innovation Paradox (see e.g. Dosi et al., 2006).

The response of many (European) research institutions to this Innovation Paradox has been to establish a TTO with a mission to stimulate, coordinate and support the commercialization of scientific and research results. A study by De Cleyn et al. (2010) indicated that many TTOs in Europe were established in the late 1990s. These TTOs are (potentially) important actors in the Triple Helix Model, which describes the relationship between academia, business and government. The concept of the Triple Helix, referring to university–industry–government relationships, was initiated in the 1990s by Etzkowitz (1993). The concept reflects on the (increasing) importance of knowledge in the triadic relationship between research institutions, industry and government. The same study supports the statement that most research institutions around the globe have limited experience (less than 25 years) of organizing structural knowledge transfer via TTOs such as academic spin-offs or technology licensing.

Academic Entrepreneurship and the Triple Helix

The European Innovation Paradox, together with changes in society, has forced research institutions to rethink their role and contribution to business and society.

As mentioned earlier, one response has been the setting-up of TTOs and intensifying the attention paid to and support of academic entrepreneurship within these institutions. TTOs play an active role in commercializing Public Research Organization (PRO) research by identifying, protecting, marketing and licensing IP developed by researchers. However, in analysing the impact of such TTO activity, studies have focused more on the effectiveness of technological diffusion that used licensing rather than spin-offs as a commercialization mechanism (Siegel et al., 2003a). There are fewer studies focusing on the impact of spin-off activity and TTO engagement with start-ups and SMEs. The main outcomes of studies by Lockett and Wright (2005) and Powers and McDougall (2005) relate to the importance of TTO size and experience in relation to spin-off activity at research institutions.

Increased engagement by research institutions in the support and promotion of entrepreneurial activities is not only an internal reflex. These institutions have become increasingly aware of the context in which they operate, no longer functioning as 'ivory towers' in which new knowledge is produced and disseminated through scientific publications and education, but as important organizations embedded in a local and global ecosystem where different actors interact (the Triple Helix Model). They can play an important role in the dynamics of a region, given their ability to create knowledge, attract companies to relocate close to them and foster job creation via spin-offs (Jones-Evans et al., 1999). Additionally, PRO inventions are an important source of knowledge spillovers (Di Gregorio and Shane, 2003), potentially benefitting numerous stakeholders in a region.

Technology and Knowledge Transfer

As mentioned earlier, this book builds on prior studies that have sought to understand the models, implications and success factors of technology and knowledge transfer programs and organizations in research institutions as well as in corporate environments. TTOs play an important role in supporting the translation of academic knowledge into applications. With the help of TTOs, research institutions can potentially contribute to local and global economies at various levels: by developing new technologies and knowledge that provide solutions to the challenges faced by business and society (Etzkowitz et al., 2000); by introducing innovation and – e.g. through spin-off ventures – creating new businesses that create jobs (Jones-Evans et al., 1999); by sharing new knowledge that is beneficial to a wide range of stakeholders (other researchers, students, policymakers, companies, etc.) (Gunasekara, 2006); and even by becoming critical to an entire industry, such as pharmaceuticals, which has (at least partially) become dependent on academic research for the discovery of new drugs (Festel et al., 2011).

At a more general level, Teece (2003) has tried to understand the cost of transferring technological know-how from multinationals. Indeed, knowledge has become a key raw material for many economies. And yet, despite the rise of digital technologies, transferring knowledge from one actor to another remains challenging. Other studies have focused on technology and knowledge transfer from research institutions (e.g. Lee, 1996; Siegel et al., 2003b). In this context, a recent study by Clarysse et al. (2014) have identified a gap in the knowledge and business ecosystem, which is currently not bridged by any initiative. The flipped knowledge transfer approach (as highlighted in the contribution by De Cleyn and Gielen in this book) could help to (partially) bridge this gap.

From a more financial perspective, Bray and Lee (2000) have analysed the effect of technology and knowledge transfer on research institutions. Their study concludes that institutions may benefit more from taking equity in their spin-offs than from licensing deals. In general, equity generates a return on investment equal to licensing contracts, but equity offers the advantage of occasionally hitting the 'jackpot'. Overall, equity in spin-offs maximizes the financial return research institutions can realize from their IP.

Academic Spin-Offs as a Mechanism for Knowledge Transfer

The academic spin-off (ASO) is one of the more visible mechanisms for knowledge transfer. These new ventures are set up by research institutions in order to translate specific research results (new knowledge and/or technologies) into (commercial) application. ASOs can be established by (former) academic researchers, external entrepreneurs working with PRO technologies, and/or students (student-led spin-offs).

ASOs have specific characteristics that distinguish them from other (technology-based) start-ups. First, ASOs face the same market risk as any other start-ups, i.e. they need to find 'product–market fit' (by selling products and/or services that the market is willing to pay for). Secondly, they face a technological risk associated with the uncertain technology and subsequent product development risk. Developing new products and services based on the latest technologies always carries an inherent risk, because these technologies may not be fully understood (yet). These two risks, common to almost every technology-based start-up, are supplemented by a third risk stemming from the non-commercial background from which the ASO originates. The founders of ASOs often have an academic track record, but lack experience in a commercial environment.

Because of these risks and the absence of a prior track record at venture level (De Coster and Butler, 2005), ASOs often have difficulties in obtaining sufficient resources including but not limited to finance. Furthermore, the absence of a prior track record and an ability to objectively assess a new venture's value and potential results in low legitimacy levels. The legitimacy of an ASO stems from the founders' prior experience and knowledge (human capital), its origins (the PRO's reputation reflects on the ASO) and the relationships the founder has built up during prior activities (social capital).

Although in no sense exclusively, this book emphasizes the role ASOs (can) play in knowledge transfer activities at research institutions around Europe and the challenges they face. In this regard, various chapters focus on how to create a framework that stimulates the creation of ASOs, how ASOs are part of an entire ecosystem around a PRO, how the ASO funding challenge can be tackled and how student-led spin-offs can contribute to an effective knowledge transfer policy.

Clusters of Innovation and Economic Growth

Last but not least, this book is embedded in the concept of (global) clusters of innovation. In recent decades, it has become clear that start-ups are a major force in creating and driving innovation to the market, thereby generating economic vitality for entire regions (Engel, 2014). Attracting start-ups and creating a fertile breeding ground for them to flourish, however, requires a mix of ingredients that can only be provided at a regional scale, surpassing local capacities. AnnaLee Saxenian's seminal work investigated the differences between the Boston and Silicon Valley approaches and eventually predicted the latter's dominance (Saxenian, 1994). Her work has spurred more research in this domain and, as various regions around the globe have experimented with creating successful innovation hubs or clusters, cluster thinking has come to dominate the discussion and content of regional economic development policy.

Clusters offer many advantages to local actors: physical proximity enables economies of scale and scope, easier access to information (including the latest academic knowledge and newly developed technologies from research institutions), proximity to specialized suppliers and customers, and reduced transaction costs, among others. In real 'clusters of innovation', intense concentrations of specific industries emerge, resulting from an ongoing process of new start-up creation and fast commercialization and adoption of new technologies (Engel, 2014; Engel and del-Palacio, 2009). These clusters of innovation are characterized by mobile assets (e.g. money, people and information, including know-how and IP). Additionally, an entire service industry develops around these new start-ups and maturing businesses.

The concept of clusters of innovation is closely related to the Triple Helix Model, where interactions between academia, industry and government play a central role. Indeed, many clusters of innovation, including the textbook example of Silicon Valley, have at some point been initiated or accelerated by some form of government intervention. In order to fuel the creation of a cluster of innovation, several authors argue that especially the interactions between industry and academia are of critical importance. In this sense, research institutions and their TTOs play a crucial role in unlocking new knowledge and research outcomes for start-ups and other actors in a regional cluster.

WHY THIS BOOK?

The objective of this book is to provide an overview of the recent challenges research institutions in Europe are facing in technology and knowledge transfer, and to share some of the latest solutions different institutions have developed to achieve more effective and successful knowledge transfer. Most research institutions now have between 10 and 25 years of experience of knowledge and technology transfer using a structured and organized approach (often through TTOs). And yet most of them are still searching for better mechanisms to organize these activities and obtain maximal results. Major progress has been made in recent decades, in practice as well as from a research perspective. As a result, the authors felt it was time to recap what has been learnt and to look ahead at solutions for current imperfections and new challenges that will emerge. It is in this spirit that the authors have assembled case studies that illustrate the impact of knowledge transfer activities on specific regions and organizations. It is our hope that these examples will assist others in recognizing, supporting and enhancing opportunities to develop or improve knowledge transfer activities in their own milieus.

At the conclusion of this volume, we contextualize the key insights from these cases, and map them into the main trends and challenges in knowledge transfer from previous years and for years to come. This can lead to a better understanding of the general phenomenon of knowledge transfer from research institutions and how to optimize it in the future.

REFERENCES

Bray, M.J. and J.N. Lee (2000), 'University revenues from technology transfer: Licensing fees vs. equity positions', *Journal of Business Venturing*, **15** (5–6): 385–92.

Clarysse, B., N. Moray and A. Heirman (2002), 'Transferring technology by spinning off ventures: Towards an empirically based understanding of the spin-off process', Working Paper, Universiteit Gent, Faculteit Economie en Bedrijfskunde, **131**: 1–32.

Clarysse, B., M. Wright, J. Bruneel and A. Mahajan (2014), 'Creating value in ecosystems: Crossing the chasm between knowledge and business ecosystems', *Research Policy*, **43** (7): 1164–76.

De Cleyn, S.H., R. Tietz, J. Braet and M. Schefczyck (2010), *Report on the status of academic entrepreneurship in Europe: 1985–2008* (Puurs: Unibook).

De Coster, R. and C. Butler (2005), 'Assessment of proposals for new technology ventures in the UK: Characteristics of university spin-off companies', *Technovation*, **25** (5): 535–43.

Di Gregorio, D. and S. Shane (2003), 'Why do some universities generate more start-ups than others?', *Research Policy*, **32** (2): 209–27.

Dosi, G., P. Llerena and M.S. Labini (2006), 'The relationships between science, technologies and their industrial exploitation: An illustration through the myths and realities of the so-called "European Paradox"', *Research Policy*, **35** (10): 1450–64.

Engel, J.S. (2014), *Global clusters of innovation: Entrepreneurial engines of economic growth around the world* (Cheltenham: Edward Elgar).

Engel, J.S. and I. del-Palacio (2009), 'Global networks of clusters of innovation: Accelerating the innovation process', *Business Horizons*, **52** (5): 493–503.

Etzkowitz, H. (1993), 'Enterprises from science: The origins of science-based regional economic development', *Minerva*, **31** (3): 326–60.

Etzkowitz, H. (1998), 'The norms of entrepreneurial science: Cognitive effects of the new university – industry linkages', *Research Policy*, **27** (8): 823–33.

Etzkowitz, H., A. Webster, C. Gebhardtand and B.R.C. Terra (2000), 'The future of the university and the university of the future: Evolution of ivory tower to entrepreneurial paradigm', *Research Policy*, **29** (2), 313–30.

European Commission (1995), *Green Paper on Innovation* (Brussels, Belgium: European Commission).

Federation of Small Businesses (FSB) (2014), 'Statistics', http://www.fsb.org.uk/stats (accessed 27 April 2016).

Festel, G., S.H. De Cleyn, R. Boutellier and J. Braet (2011), 'Optimizing the R&D process using spin-outs: Case studies from the pharmaceutical industry', *Research-Technology Management*, **54** (1): 32–41.
Gunasekara, C. (2006), 'Reframing the role of universities in the development of regional innovation systems', *The Journal of Technology Transfer*, **31** (1): 101–13.
Guston, D.H. (1999), 'Stabilizing the boundary between US politics and science: The role of the Office of Technology Transfer as a boundary organization', *Social Studies of Science*, **29** (1): 87–111.
International Finance Corporation (IFC) (2012), 'IFC and small and medium enterprises', http://www.ifc.org/wps/wcm/connect/277d1680486a831abec2fff995bd23db/AM11IFC+IssueBrief_SME.pdf?MOD=AJPERES (accessed 27 April 2016).
Jones-Evans, D., M. Klofsten, E. Andersson and D. Pandya (1999), 'Creating a bridge between university and industry in small European countries: The role of the Industrial Liaison Office', *R&D Management*, **29** (1): 47–56.
Klofsten, M. and D. Jones-Evans (2000), 'Comparing academic entrepreneurship in Europe – the case of Sweden and Ireland', *Small Business Economics*, **14** (4): 299–309.
Lee, Y.S. (1996), '"Technology transfer" and the research university: a search for the boundaries of university-industry collaboration', *Research Policy*, **25** (6): 843–63.
Lockett, A. and M. Wright (2005), 'Resources, capabilities, risk capital and the creation of university spin-out companies', *Research Policy*, **34** (7): 1043–57.
Nunes, M.B., F. Annansingh, B. Eaglestone and R. Wakefield (2006), 'Knowledge management issues in knowledge – intensive SMEs', *Journal of Documentation*, **62** (1): 101–19.
Powers, J.B. and P.P. McDougall (2005), 'University start-up formation and technology licensing with firms that go public: A resource-based view of academic entrepreneurship', *Journal of Business Venturing*, **20** (3): 291–311.
Santoro, M.D. and P.E. Bierly (2006), 'Facilitators of knowledge transfer in university-industry collaborations: A knowledge-based perspective', *IEEE Transactions on Engineering Management*, **53** (4): 495–507.
Saxenian, A.L. (1994), *Regional advantage: Culture and competition in Silicon Valley and Route 128* (Cambridge, MA: Harvard University Press).
Siegel, D.S., D. Waldman and A.N. Link (2003a), 'Assessing the impact of organizational practices on the productivity of University Technology Transfer Offices: An exploratory study', *Research Policy*, **32** (1): 27–48.
Siegel, D. S., P. Westhead and M. Wright (2003b), 'Assessing the impact of university science parks on research productivity: Exploratory firm-level evidence from the United Kingdom', *International Journal of Industrial Organization*, **21** (9): 1357–69.
Singh, R.K., S.K. Garg and S.G. Deshmukh (2010), 'The competitiveness of SMEs in a globalized economy: Observations from China and India', *Management Research Review*, **33** (1): 54–65.
Teece, D.J. (2003), 'Capturing value from knowledge assets: The new economy, markets for know-how, and intangible assets', in D.J. Teece, *Essays in technology management and policy* (Singapore: World Scientific).
Teece, D.J. (1998), 'Capturing value from knowledge assets: The new economy,

markets for know-how, and intangible assets', *California Management Review*, **40** (3): 55–79.
Tijssen, R.J.W. and E. van Wijk (1999), 'In search of the European Paradox: An international comparison of Europe's scientific performance and knowledge flows in information and communication technologies research', *Research Policy*, **28** (5): 519–43.

Summary of contributions

Sven H. De Cleyn and Gunter Festel

Some important hurdles hamper the commercialization of scientific knowledge in Europe. They include the well-known technology transfer gap between academic research at universities or research institutions and industrial application. This gap between academic research and the commercialization of new knowledge can be addressed by spin-off ventures, which translate new knowledge and technologies into market relevant applications. Spin-offs are usually more flexible and faster in doing this than established companies, given their lean structure and absence of any prior track record. They are an important mechanism to address customer needs and bring new products and services to the market. In this regard, they have been demonstrated to foster the development of regional clusters and accelerate the economic growth of a region, as several contributions in this book confirm.

Against this backdrop, universities are increasingly recognizing the importance of start-ups. Since many academic researchers neither have the knowledge nor the experience (or ambition) to commercialize their academic research results, universities and other public research organizations are steadily setting up a variety of support mechanisms to facilitate technology transfer, such as technology transfer offices (TTOs), investment funds, incubator facilities and coaching. Most TTOs recognize spin-offs as an interesting method of technology transfer and seek to support scientists in their entrepreneurial efforts. However, Europe has traditionally performed poorly in translating new (academic) knowledge into economic and social value, despite its broader scientific base when compared to, for instance, the US. This phenomenon has become known as the European Innovation Paradox, because until now Europe has failed to convert this advantage into a strong contribution to economic development and innovation. This is attributed to the more competitive and flexible American academic environment and legislative system.

The risks associated with spin-offs are twofold: (i) they face a general market risk, which any start-up has to overcome; and (ii) their (often) high-tech nature exposes them to a technological risk, stemming from the complex technology and/or product development process. In the case of academic spin-offs, these two risks are complemented by a third: their non-commercial background. Given these risks and the high level of financing required to cover the cost of product development, the available options for funding academic spin-offs are limited. In addition to capital, these spin-offs often also lack business know-how because the founders are mainly research-orientated scientists. Therefore, they also rely heavily on operational assistance in order to be successful. The earlier a working relationship between founders and investors begins in the development process, the less likely conflicts regarding goals or tasks will occur. Such conflicts often lead to the exit of investors and entrepreneurs.

A BRIEF OVERVIEW OF INSIGHTS IN THE COMING CHAPTERS

The contributors to this book have been carefully selected for their specific expertise in technology transfer and academic spin-offs in a broad European context. Their contributions examine diverse aspects of the spin-off phenomenon and take a closer look through different lenses.

The first three chapters focus on ways to shape an ecosystem that is receptive to technology transfer and stimulates the creation of academic spin-offs. As the different contributions show, the right framework conditions play a major role in the success (or failure) of building an entrepreneurial university and innovation ecosystem.

In Chapter 1, Marco Cantamessa from the Politecnico di Torino (Italy) describes the dual mission of I3P, his university's business incubator, in technology transfer and start-up ecosystem development. He shows how a successful incubation initiative can be developed, provided that stakeholders are able to define a sustainable business model that fits with regional innovation policy, and for which internal operations are given the right incentive set. When a university incubator acts as a public institution addressing a market failure, there is a strong risk of creating a self-serving organization with a business model that is not properly aligned to its strategic objectives. After a brief presentation of the historical and territorial environment in which I3P operates and the results achieved so far, the chapter discusses key elements such as (i) ownership, governance and strategic objectives, (ii) vertical and horizontal integration choices and (iii) operational aspects. Moreover, the core competence of an incubator is to provide a specific kind of consultancy service to start-ups. This activity relies on human knowledge and experience, which makes it crucial to develop an appropriate human resources management strategy including recruitment, professional development, responsibility allocation and individual incentives.

In Chapter 2, Elco van Burg from VU University, Amsterdam; Isabelle Reymen and Georges Romme from Eindhoven University of Technology (the Netherlands); and Victor Gilsing from University of Antwerp and Tilburg University, describe strategies and design processes for new venture units in complex contexts (universities). The fact that these 'incubation' units are separately organized makes the design process and interaction with the complex context an intriguing yet little-understood endeavor. Their chapter looks at how the design process and context interact. Key contextual characteristics are the degree of loose coupling of organizational units (near-decomposability), the hierarchical nature of the organization, and the interactions between designers and external stakeholders. Drawing on a comparative case study of three universities in Spain and the Netherlands that have designed new venture units to commercialize university inventions, they describe how organization designers use three strategies in the design process: off-line reasoning and planning, feedback-driven learning and associative reasoning by way of analogies. Their findings suggest that associative reasoning is the primary design strategy in contexts characterized by a high degree of near-decomposability and hierarchy. In addition, feedback-driven learning appears to be necessary in anchoring the designs to make the new venture unit viable over time. As such, this study contributes to innovation and corporate entrepreneurship literature and to organization design literature by describing the relationship between context characteristics and different design strategies as well as specifying the contributions of these strategies to the performance of the design process in these contexts.

In Chapter 3, Matthias Mrozewski, Agnes von Matuschka, Jan Kratzer and Gunter Festel present an entrepreneurial university in an entrepreneurial city. Berlin is considered as one of the most entrepreneurship-friendly cities in Germany, and one of the major players is Technical University Berlin (TU Berlin), also one of the country's oldest and largest technical universities. TU Berlin has a long track record in the commercialization of innovation and technology transfer and is regarded as a pioneer in entrepreneurship support. The university actively strives to embed entrepreneurship at every level of its organization and to leverage the huge potential for commercializing inventions and research results. TU Berlin concentrates its competences in research, education and practical entrepreneurship support within the Center for Entrepreneurship (CfE), a platform for entrepreneurial students and scientists, potential investors, business angels and other entrepreneurship support systems. CfE services are coordinated and executed by the university's Chair of Entrepreneurship and

Innovation Management and its Founder's Service. This close collaboration guarantees that students receive the right balance of academically demanding lectures as well as practically oriented workshops and seminars. In recent years, a vibrant network of founders, companies, business angels, venture capital investors and industry experts has emerged.

After focusing on the ecosystem, the next three contributions highlight successful examples of support programs. Each in their own unique way has implemented intelligent programs to develop entrepreneurial skills among researchers and to prepare future entrepreneurs for the challenging task of building technology-based spin-offs.

In Chapter 4, Helmut Schönenberger from TU Münich (Germany) focuses on the process needed to launch academic spin-offs capable of quickly developing into global players. In 2013, TU Münich's KICKSTART program (an incubation program developed by UnternehmerTUM) founded more than 20 scalable high-tech start-ups. This chapter outlines the historical context in which UnternehmerTUM emerged, how the program was designed for maximum impact and what the current challenges are. Important success factors include peer learning, creating a vibrant network with existing companies (corporate networks) and a competitive intake process for teams rather than individuals. Despite its success, the KICKSTART program is still trying to push the boundaries in order to become self-reinforcing (i.e. generating enough spin-offs that become global players to create a cluster dynamic for even more scalable start-ups).

Chapter 5, by Magnus Klofsten from Linköping University (Sweden) and Erik Lundmark from Macquarie University, Sydney, Australia, digs deeper into the coaching of academic spin-offs. Their ENP program (Entrepreneurship and New Business Program) encompasses more than 20 years of experience in coaching and training new generations of spin-off founders. Over the years, more than 1,500 participants have followed the program and founded in excess of 500 new spin-offs in Sweden and elsewhere in Europe. The program combines business plan preparation, coaching, mentorship, workshops by experts and access to networks. The ENP story illustrates that a new spin-off venture is not the only positive outcome. The program also enriches the skills and mindsets of researchers so that they approach future research differently and are more receptive to research commercialization.

In Chapter 6, Julia Bauer, Matthias Keckl, Thorsten Lambertus and Björn Schmalfuß explain how Intrapreneurship Labs are stimulating a more entrepreneurial mindset among technological competence owners at the Fraunhofer-Gesellschaft (Germany). Despite being Europe's largest application-oriented research organization, the number of spin-offs from

Fraunhofer institutes is stagnating. Because of the large number of industry projects, Fraunhofer researchers have limited time to work on their own ideas; furthermore, in the current (positive) economic climate, industry partners offer them well-paid and stable jobs. As a result, the directors of Fraunhofer institutes have little incentive to support spin-offs. The aim of the 'Fraunhofer Fosters Intrapreneurship' (FFI) initiative is to change this. FFI is building Intrapreneurship Labs in the institutes to nurture a sustainable entrepreneurial mindset among Fraunhofer researchers and eventually increase the number of spin-offs by employees.

Having created the right conditions (ecosystem and framework) and support programs, many academic spin-offs still face a major funding gap. Two case studies zoom in on the funding challenges of young academic spin-offs and offer creative solutions to address them.

In Chapter 7, by Ulrich Mahr and Florian Kirschenhofer, Mahr from Max-Planck-Innovation (Germany) describes his organization's approach to addressing the 'innovation gap'. In the early 1980s, the European Union (EU) recognized the need to support entrepreneurial projects and small and medium-sized companies (SMEs) in order to develop regions and drive economic growth. The EU's solution was to provide knowledge-based services, complementary to the then-existing models of solely supplying affordable office space and services. These complementary services included training and coaching as well as advice on technical and business development challenges and access to relevant networks. Today, more than 150 Business Innovation Centers (BICs) or incubators exist Europe-wide, in addition to those meanwhile set up by European member states, regions, companies, private persons, universities and research units. Technology transfer organizations like Max-Planck-Innovation GmbH are now introducing new concepts and initiatives that include providing capital for the incubation phase as well as the later start-up phase. Capital typically comes from both public and private sources coordinated by the incubator. After the incubation phase, projects should reach a development stage that matches the demands of third party investors or industrial partners. Two examples of new incubation concepts – the Life Science Inkubator in Bonn and Dresden, and the IT Inkubator in Saarbrücken – detail the components needed to efficiently address the 'innovation gap'.

In Chapter 8, Michael Brandkamp from High-Tech Gründerfonds (Germany) reflects on the investment side of the spin-off journey. Starting from the German seed capital market, this chapter diagnoses the market failure of early stage funding. Especially at the seed stage, obtaining funding is particularly challenging given information asymmetries, limited liquidity for investors and high uncertainty at different levels (team, market, technology). Building on current literature and practice on the seed stage funding gap, this case study explains how High-Tech Gründerfonds aims to provide a sustainable solution, specifically in the context of academic spin-offs, and, from a broader perspective, new technology-based firms. As this chapter suggests, funding is important for the survival of academic spin-offs, but is not the only key to success.

The next three contributions add new flavor to the academic spin-off and technology transfer field by presenting innovative tools that could further stimulate the spin-off process and make it more successful.

In Chapter 9, Gunter Festel from Festel Capital (Switzerland), University of Basel (Switzerland) and TU Berlin (Germany) explains how the founding angel (FA) investment model is emerging to support early stage high-tech start-ups. FAs engage at a very early stage with business idea discussions, long before the arrival of business angels (BAs) and venture capitalists (VCs). Together with scientists, FAs found high-tech start-up companies and successfully commercialize academic research results. They complement the scientific knowledge of the team with their business expertise. Besides initial funding in the (pre-) seed phase, FAs are operationally engaged, bringing expertise from previous, successful start-up projects. Because of their early and much more operational engagement, the role of the FA is more as founder and entrepreneur than investor, complementing the later engagements of BAs and VCs. Based on FA engagements, significant additional investments can be raised and the profitability of invested money is generally high, with annual returns of more than 100% in successful engagements.

In Chapter 10, Sven H. De Cleyn and Frank Gielen from iMinds (Belgium) describe a new way to deal with technology transfer, in which a more market-driven approach plays an important role. Traditional technology transfer starts from academic research, for which new applications are then developed. The Flipped Knowledge Transfer approach described in this chapter uses the needs of start-ups and SMEs as the starting point for knowledge transfer, leading to more demand-driven knowledge transfer initiatives. This approach enriches the portfolio of mechanisms that can be used to successfully transform research results into market relevant applications. Furthermore, it enriches future research and speeds up time-to-market for academic output.

In Chapter 11, Mia L. Justesen from University College Nordjylland and Rajiv V. Basaiawmoit, Flemming K. Fink and Kirstine V. Moltzen from Aarhus University (Denmark) take us on a journey through student spin-offs and their role in stimulating the 'entrepreneurial university'. The university's Student Incubator approach is research-based and grounded in effectuation logic, self-efficacy and the push-method. The focus is not only on academic knowledge and business skills, but also on the individual(s) involved, their means and ability to act upon opportunities.

By comparing the European innovation landscape with that of the US, the last contribution offers the reader an international perspective.

In Chapter 12, Hervé Lebret from Innogrants (EPFL, Switzerland) broadens the scope to compare the spin-off experience in Europe and the US. The chapter starts by confirming the leading role of the US in terms of innovation and entrepreneurship, not only in terms of business, but also the higher degree of entrepreneurship found in the country's academic institutions. Subsequently, the chapter digs deeper into the reasons behind these differences, mainly attributing them to a different cultural approach towards entrepreneurship. Given the larger talent pool in the US (and especially in Silicon Valley), this chapter explains why Europe should not strive to reproduce a second Silicon Valley, but rather to celebrate its own entrepreneurs and provide role models that inspire a new generation of research-driven spin-offs and start-ups.

We hope that all 12 contributions will contribute to the reader's understanding of the current challenges facing the creation of successful academic spin-offs and fruitful technology transfer activities. Not only do they provide a critical overview, but they also share innovative approaches and solutions for building a stronger framework to nurture entrepreneurship in Europe.

PART 1

Shaping the ecosystem

1. I3P as university business incubator – a dual mission in technology transfer and start-up ecosystem development

Marco Cantamessa

1.1 UNIVERSITY INCUBATORS IN REGIONAL INNOVATION SYSTEMS

In the literature dealing with innovation systems, policies and instruments designed to stimulate entrepreneurship and foster new venture creation feature strongly (Hekkert et al., 2007). In reality, a key avenue for the innovation process seems to pass through the creation and growth of new, innovative firms. Start-ups can influence the innovation performance of a territory through their growth, as well as through the positive impact they can have on incumbent firms through trade in new goods and services, knowledge spillovers and competition (Erkko and Yli-Renko, 1998).

The nurturing of start-ups is therefore generally considered a key element in regional innovation policy. Scholars, consultants and policy-makers have long viewed the creation of an environment conducive to new and successful start-ups as the 'holy grail' of a regional innovation system, and it is very common for those engaged in this field to share the dream of creating 'their own Silicon Valley'.

Given this aim, in recent decades incubators have been seen as useful tools for stimulating the process by which new technology-based companies are created and matured. Over the years, start-up incubation has been subject to a variety of models, including public vs. private ownership, non-profit vs. for-profit orientation, narrow vs. generalist sectorial specialization and multiannual incubation programs vs. very short 'acceleration' initiatives (Peters et al., 2004).

Among the actors developing incubation programs, universities have always played a key role. This has occurred because of the increasing attention paid by academic institutions towards their 'third mission' (i.e. economic development, alongside their traditional orientation towards research and education (Zomer and Benneworth, 2011)) and towards the associated emergence of the so-called 'entrepreneurial university' (OECD-EC, 2012). In this context, universities have approached start-up incubation with the dual aim of bringing research results and technical competencies to the market, and of providing practical entrepreneurial opportunities for their academic staff and students. In an 'entrepreneurial university', entrepreneurship can therefore be viewed as a practical avenue for technology transfer, as a subject to be researched and taught, and as a general mindset that permeates the institution at a more general level (Cantamessa, 2015).

University incubation programs generally do not exist in isolation. They are part of the broader strategy with which an academic institution relates to its social and economic environment. Therefore, they are an integral part of the so-called 'Triple Helix' model of innovation (Etzkowitz and Leydesdorff, 2000) that links the actions of academia, industry and policy-makers in what hopefully becomes a complementary and synergistic action.

As such, managing a university incubation program is complex. Aside from the inner complexity of carrying out its core business of translating business ideas into successful companies, it must tackle the outward complexity of effectively linking with the academic milieu and with the regional socioeconomic environment, exploiting this potential and co-evolving with it.

The aim of this chapter is to shed some light on the multifaceted nature of university incubation, starting with a single data point represented by the I3P incubator of Politecnico di Torino in Italy. A single case study is obviously not sufficient to provide a robust empirical grounding that will lead to either the exploration of hypotheses to be tested or – even more so – to the confirmation of conclusions. Therefore, this chapter limits its ambition to providing anecdotal evidence that may stimulate further debate and research on the topic of academic incubation programs. The following section will provide a brief description of I3P's history, followed by an analysis of the main aspects that characterize a university incubator from an economic perspective, and attempt to draw some managerial implications.

1.2 THE EXPERIENCE OF I3P

1.2.1 History and Profile

I3P is the technology incubator of Politecnico di Torino, Italy's second largest technical university, ranked sixteenth in Europe by the QS2014 ranking for technical universities. It has 33,000 undergraduate and graduate students (of which 700 are at doctoral level), 800 lecturers and an annual budget of 260 million euros.

I3P is located in the city of Turin, in the northwestern part of the country. Turin is a medium-sized city with 870,000 inhabitants (1.7 million including its metropolitan area) and a long industrial tradition. The city's industrial focus has progressively moved from textiles in the nineteenth century to automotive and aerospace in the twentieth century. At the end of the 1990s, the Municipality initiated its first Strategic Plan, with the objective of leading the transformation of the city and its surroundings beyond the traditional dominance of the automotive industry. Turin at that time was considered 'Italy's Detroit', but with the difference that a single carmaker (Fiat) was the dominant player, with links to the city that were seen to be dwindling, along with a perceived lack of competitiveness in a globalizing world. One of the chapters in the Strategic Plan concerned the nurturing of new high-tech businesses, and the establishment of an incubator was therefore included in the plan. The I3P incubator was set up in 1999 as an independent non-profit-making company owned by the Politecnico di Torino and the local public authorities (the Municipality, the Provincial and Regional Governments, the Chamber of Commerce), each with equal voting rights.

This shared ownership and the strategic orientation of its shareholders led I3P to envisage a broader mission than simply supporting university spin-offs, and to define itself as an actor aiming to create qualified jobs through the growth of high-tech start-ups. Therefore, from the outset I3P started scouting for entrepreneurial ideas not only within the university, but also across the region. Currently, each stream accounts for around half of all applications coming to I3P. The former includes ideas derived from research projects (usually with a strong scientific grounding but an unclear business applicability) as well as simpler business ideas coming from students. The latter mostly consists of applications from entrepreneurs living in the area, many of whom are alumni of the university, or attracted to Turin and to the Politecnico campus because of the technical and industrial competencies they find there. Additional applications come from companies wishing to create corporate spin-offs, usually in an attempt to extract value from non-core projects that have been developed up to a preliminary stage and to provide entrepreneurial incentives to staff in charge of these projects.

This broader mission and orientation had a second implication. It was decided not to specialize the incubator in a specific technology or industry, but to cover all the fields in which Politecnico di Torino could provide competencies (effectively excluding only biotechnology and pharmaceuticals).

The first corollary of this generalist orientation was not to set up shared laboratory facilities. Start-ups are invited to set up their own labs as long as the investment is reasonable, and to partner with academic departments when capital-intensive equipment is required. This decision allowed the incubator to reduce its capital and operating expenditures and – as a consequence – to stimulate exchanges between start-ups, and between start-ups and academic departments. I3P recognized that a generalist strategy ran the risk of de-focalization. To counter this, it aimed at developing a large-scale incubation program, which would allow the specialization of its team of coaches[1] in industry-specific 'practices', following the model of management consultancy companies. Currently, I3P is organized into four practices: IT/Internet, CleanTech, MedTech and Automation. In other words, it was thought that a large incubator made up of multiple practices would simultaneously offer the advantages of both specialized and generalist orientations. Each practice involves senior coaches who have significant experience in supporting start-ups, junior consultants who provide most of the contact hours with the start-ups, and junior trainees who support the most labor-intensive activities, such as collecting market information and setting up economic and financial models and cap tables. In order to create a flexible organization and to imbue it with greater entrepreneurialism, I3P coaches are mostly freelancers employed by the incubator on medium-term contracts and for a substantial portion of their working time, although they do not work exclusively for the incubator. This gives them the freedom to work with external clients and contribute to the diffusion of a 'start-up culture' across the territory.

I3P's close relationship with its public shareholders allows it to quickly launch initiatives and respond to trends while leveraging its cumulated experience. For instance, when in 2012 the Municipality launched an initiative to support start-ups in the field of social innovation, it did not have to establish a new entity. Rather, it worked with I3P (which provided entrepreneurial know-how and the enabling technology, along with its sister incubator at the University of Turin) and with other partners who contributed specific experience in the field of social innovation.

As a publicly owned entity, I3P constantly runs the risk of passively relying on public funding, thus impairing the sustainability of its business model and – most of all – losing its entrepreneurial streak, which would contradict its mission and damage its credibility among start-up founders. Therefore, the business model of I3P has been subject to a number of experiments, until a satisfactory solution was found in 2008. This solution stemmed from the recognition that, given the time required for start-ups to grow and the absence of private donors, there was a market failure that justified public contribution. However, it was also decided that this contribution, drawn from the European Social Fund (ESF) budget of the Piedmont Region, should not cover all the incubator's expenses and should also include a correct set of incentives similar to those found in a well-functioning market. To address the former criterion, a 35% revenue target coming from tenant companies was identified as a reasonable objective. To address the latter criterion, the funding scheme was linked to operational results and not only to the reporting of expenses – as required by ESF regulations. The mechanism works roughly like this: each time a new start-up is incorporated, I3P is authorized to claim reimbursement of expenses up to the predefined 'standard cost' of launching a new firm. Being paid according to the number of start-ups incorporated creates a clear incentive since – if the incubator should stop launching new companies – this would end both public funding and revenues derived from service fees paid by tenants. Given that the number of start-ups created is the main revenue driver of the incubator, this has spontaneously led to an MBO (Management By Objectives) mechanism, in which each of the incubator's practices is evaluated according to the start-up creation process (i.e. number of applications received, number of start-up projects worked upon, and actual number of start-ups incorporated).

An alternative business model, based on taking equity from start-ups, was considered but not adopted, mainly because of the time required to grow a start-up on the Italian market. Moreover, this would have biased the activity of the incubator towards 'quick wins' and away from projects with a stronger technology base, which require a longer time-to-market. Finally, because public institutions own I3P, legislative complexity would have made the management of these equity stakes fraught with compliance risk and administrative overheads.

1.2.2 The Incubation Process

The incubation process followed by I3P is relatively standard and based on four main activities.

Scouting concerns finding new entrepreneurs and business ideas. One key avenue for scouting is via person-to-person and word-of-mouth contacts within the Politecnico di Torino and its labs, as well as in the local territory. Alongside this, I3P organizes a significant number of events, such as Start-up Weekends, thematic *hackatons* (many sponsored by large companies) and an annual Business Plan Competition. This competition is part of a nationwide program run by PNICube, the association of Italian university incubators. In 2011, I3P also launched an acceleration program called Treatabit, aimed at fast-tracking internet start-ups. Altogether, these activities lead to more than 300 applications per year to the incubation program. All applications lead to a direct contact between incubator coaches and entrepreneurs and to the establishment of a preliminary consultancy path. Following these early steps, it becomes progressively clear whether a project should be supported or abandoned.

Consultancy covers all the phases starting from initial screening and throughout the mentoring and coaching activities that allow business ideas to become well-structured projects (more than 100 per year) and – later – start-ups (about 20 per year). I3P focuses its consultancy on strategy and – depending on the type of start-up – follows either a traditional approach based on business modeling and business planning, or the recent Lean Start-up approach (Ries, 2011). Besides strategic consultancy, projects and start-ups benefit from the technical expertise of Politecnico di Torino and other technology partners. Moreover, entrepreneurs receive specialized consultancy in administrative, legal and intellectual property matters from professional services firms that I3P has selected as partners and that provide high-quality services at modest rates. From the perspective of these partner firms, this is a way to gain new customers and to create competencies and a reputation on the start-up market. From the perspective of I3P, these partnerships allow it to provide cost-effective and high-quality services to its start-ups, while at the same time stimulating existing local actors into contributing to the creation of a richer start-up ecosystem.

Teambuilding and fundraising concerns projects that are mature enough to attract talent and capital, and continues throughout the growth of the start-up. Often, the initial entrepreneurial team is not strong enough, or lacks either technical or managerial competencies. The incubator therefore uses its network to attract key personnel either as co-founders or as employees for the nascent start-up. Once this is accomplished, the start-up can be presented to equity investors for an initial funding round. Typical investors include business angels and venture capital funds, together with local companies willing to operate an informal style of corporate venturing. I3P specializes in seed financing, with rounds ranging from 50,000 euros (typical for internet start-ups) to 500,000 euros (typical for a CleanTech or MedTech initiative) and an average annual fundraising amounting to 3 million euros. I3P also supports fundraising for subsequent financing rounds, although the results remain fairly meager, given the weakness of the Italian venture capital industry, which includes few active early-stage funds. Hopefully, recent legislation that grants tax breaks to investors in start-ups and in venture capital funds will stimulate the industry and lead to a higher number of players.

I3P also supports debt fundraising, which is applicable whenever a start-up is close to achieving revenue. Banks are generally reluctant to lend money to start-ups, but I3P's reputation allows it to obtain preferential treatment for the start-ups it vouches for, based on collateral-free medium-term loans (six years, including one initial year in which the start-up must pay interest but no principal) ranging from 100,000 to 500,000 euros.

Business development deals with facilitating and supporting connections between start-ups and the market, and is probably the main activity carried out by I3P during the incubation program, which lasts three years after incorporation (plus two optional years of 'post-incubation acceleration'). Most start-ups at I3P operate in B2B markets. This makes it imperative to overcome the barriers that are often in place when a start-up attempts to interact with large corporations. Business development activity at I3P is relatively recent and started around 2010, when a few large companies showed interest in working with start-ups, along the lines of Open Innovation (Chesbrough, 2006). At present, I3P runs this activity mostly by organizing company-specific visits during which the senior management of a large company meets a selection of start-ups it might be interested in (usually eight to ten), provides feedback and chooses a subset (usually one to three) with which it may start a pilot project. These pilot projects can then lead to a fuller commercial relationship and/or to a corporate investment. Experience shows that it is imperative to have senior decision-makers taking part in these events. Middle managers often enthuse about the start-ups they see, but do not have the organizational clout to make the relationship progress at a later stage when the company slows down the process because of organizational inertia or because of internal opposition to working with start-ups.

In addition to corporate visits, I3P also promotes the participation of start-ups at international trade fairs, usually by renting a booth and sharing the costs.

1.2.3 Current Results and Future Trends

One key result of any incubation program is its capacity to generate new companies. Since its inception in 1999 and up to mid-2015, I3P has launched 175 start-ups, of which 40 are still in the incubation program. Of the remaining 135, 31 have gone out of business, 11 have been acquired, and the remaining 93 are surviving (one of which has gone through an Initial Public Offering (IPO)). Over the years, the low failure rate (under 20%) has received mixed reactions, with some stakeholders expressing satisfaction and others being somewhat perplexed. The former consider a high success rate as a positive indicator *per se* and as an indicator of a sound screening process. The latter consider the low failure rate as an indicator of a 'low risk–low reward' profile of I3P's start-ups. In principle, both perspectives are reasonable, but one must consider the business model mix of the start-ups being launched. During its early years, I3P worked mainly with start-ups characterized by a service-oriented business model, and with relatively few high-growth, product-based and high-risk start-ups, especially because at that time there was not enough equity funding available. Unsurprisingly, service-oriented businesses are less risky and account for the overall high success rate reported above. Given the progressive growth of equity funding available, the mix of start-ups has shifted towards product-based and riskier initiatives, which instead make up most of the failures.

Another important indicator of an incubation program is job creation. As shown in Figure 1.1, at the end of 2014, I3P start-ups have generated 1,400 direct jobs. By using well-known multipliers that take into account the value chains in which start-ups operate and broader economic spillovers (Moretti, 2012), it is possible to estimate the overall employment impact to range between 6,000 and 7,000 units.

The distribution of these jobs follows a Pareto distribution, with a small minority of larger companies making up the bulk of this impact. More precisely, one start-up has grown to more than 400 employees, four to five have grown to around 50, with the rest being small companies with an average

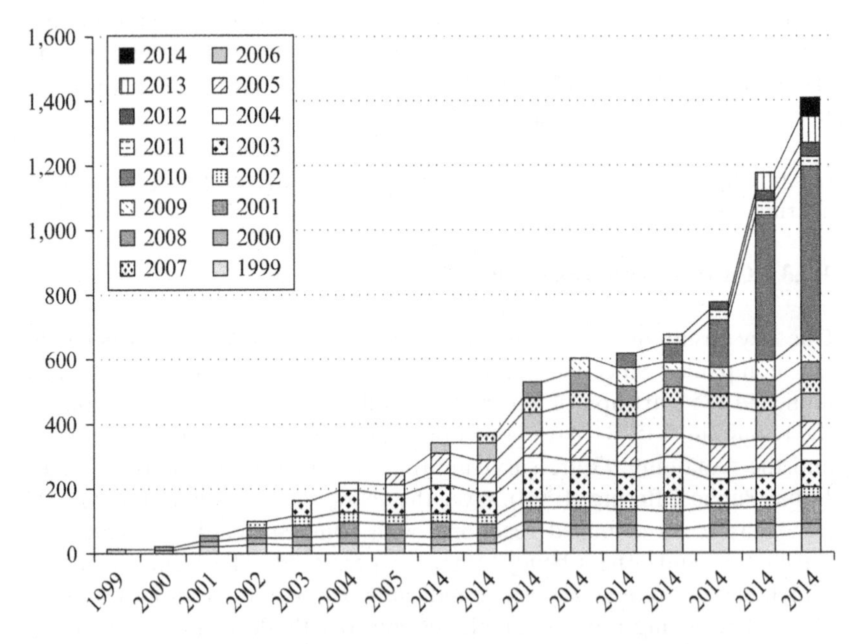

Figure 1.1 Job creation by I3P start-ups, by year of incorporation

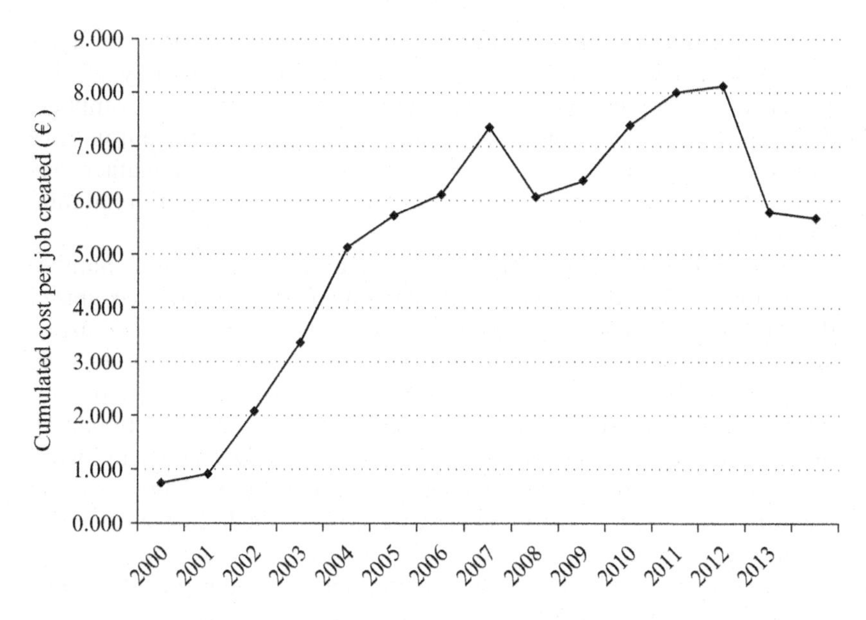

Figure 1.2 Public expenditure per job created at I3P

of five to six employees, some of which are still growing and others not. If the objective of an incubator is to generate large, world-class companies, even a sizable entity such as I3P has apparently failed to deliver. However, if the aim of an incubator is the progressive generation of a resilient stream of high value-adding jobs, while trying to generate large 'champions', then the results shown can be considered a success. Moreover, if one divides the cumulated public funding used by I3P by the number of direct jobs created, one comes up with a nearly constant value of 6,000 euros (see Figure 1.2). A one-off investment of 6,000 euros obviously constitutes a sound return for the taxpayer, given the amount of taxes and social contributions paid by each of these new employees every year.

The world of start-ups is not static. Any incubation initiative must continuously evolve in order to cater to the needs of its customers and achieve a better fit with the overall start-up ecosystem to which it belongs. In 2015, I3P was working in three main directions to shape its activities in the medium term.

The first is attracting start-ups from outside the region, leveraging the competencies available at Politecnico di Torino and, more broadly, in the surrounding territory. Turin is an interesting development hub especially for technology start-ups with some degree of manufacturing content. The city is strategically placed in Europe, provides a high number of qualified science, technology, engineering and mathematics (STEM) graduates at comparatively low salaries and offers them a good quality of life. Moreover, the city has an impressive number of manufacturing suppliers with experience of working to the stringent standards of the automotive and aerospace industries. Start-ups can connect with these manufacturers and go from a prototype to production very quickly and with minimum investment.

The second is to create tighter links with large corporations. A number of companies already working with start-ups from I3P have recognized that in order to be effective, the interaction process with start-ups must be more structured. Moreover, they have noticed that in order to successfully adopt the technologies and products being developed by these start-ups, their own products and processes must co-evolve. This calls for the development of corporate incubation programs in which entrepreneurs work closely together with incubator coaches and corporate champions and mentors. Given their lack of experience and the high cost involved in running a proprietary corporate incubation program, a number of large companies have asked I3P to host such programs along the lines of a 'white label' or 'jointly-run' model.

Finally, some limitations have been recognized in I3P's ability to foster the growth of its start-ups, which is mainly due to its public nature. In fact, by not taking equity stakes, I3P coaches cannot have a direct and active role in the governance of start-ups. In some cases, investors and entrepreneurs have asked I3P coaches to take on a more active role, assuming directorships or interim managerial positions. In order to facilitate and regulate this kind of activity, which clearly goes beyond the 'public' mission of I3P, a parallel 'acceleration program' privately run by consultants and coaches will be set up.

1.3 SOME CONSIDERATIONS FOR THE BUSINESS MODEL OF A UNIVERSITY INCUBATOR

1.3.1 The Economic Foundations of a University Incubator

It is not easy to define what a university start-up incubator is from an economic perspective. However, based on the previously discussed case study, it is possible to highlight four main economic roles and their characterizing features. As mentioned earlier, this discussion is based on a single case study and therefore represents a set of hypotheses and can by no means be considered as a validated result.

A university incubator is first a *consultancy company*, in which an experienced group of coaches and mentors provide inexperienced teams of entrepreneurs (or entrepreneurs lacking the full range of competencies) with the advice they need to set up and grow a start-up. The role of these consultants is to bring specialist knowledge and prior experience to quicken the process. Moreover, as a consultancy, the incubator can help entrepreneurs avoid costly mistakes resulting from either inexperience or typical cognitive biases (Simon et al., 2000), and thus reduce the failure rate of start-ups passing through the program. Because this activity is labor-intensive, it would not appear to generate significant economies of scale. In reality, *economies of scale* can emerge if the number of client companies is high enough to enable the structuring of a hierarchy of coaches, from junior trainees to senior consultants, thus optimizing the mix of expertise and hourly costs attached to each activity. Moreover, consulting activities clearly benefit from *learning economies*. These may arise both in domain-independent aspects (e.g. structuring a financial plan for a start-up or preparing a sales pitch) as well as industry-specific aspects (e.g. knowing the latest trends in photovoltaic plants). Therefore, a successful incubation program may require a minimum track record both in its overall activities and its activity in each industry that it chooses to operate in. Returning to the case of I3P, this dual level of experience explains the simultaneously multi-industry and large-scale approach.

A further element for consideration in the incubator's role as a consultancy is associated with the degree to which the incubator's coaches and mentors may intrude in start-ups' decision-making processes. This can range from friendly, informal advice that start-ups are under no obligation to heed, all the way to full and formal involvement in a start-up's governance. The degree of involvement has to do with the experience of the pool of coaches and mentors, with the public or private nature of the incubator, to the set of incentives it is subject to, and to the degree of potential liability the incubator is willing to bear.

A university incubator is also a *two-sided market* (Rochet and Tirole, 2006). One side of the market is made up of entrepreneurs who want to establish and grow their start-ups, while the other is made up of economic actors that want to interact with start-ups in a number of different capacities and roles. Individuals may wish to be hired; existing companies and professional services firms may look at the incubator's start-ups as potential suppliers or customers; investors will use the incubator to strengthen their deal flow.

As a two-sided market, size and variety matter significantly, thus leading to significant *economies of scale and scope*. Entrepreneurs will be attracted to an incubator if they know that the program will expose them to many, diverse and qualified economic actors. For their part, economic actors will be attracted to an incubator if they know that it hosts a large number of high-quality start-ups. Some of these economic actors may look for size and specialization, while others may be attracted by variety as well. For instance, a CleanTech venture capital fund will not be interested in a start-up developing innovative HR employee benefits. However, a large corporation operating in the CleanTech industry may find that same start-up very interesting for its own HR services.

In the context of two-sided markets, an incubator has two main distinguishing features. First, the collection of economic actors who wish to interact with start-ups is not homogeneous, and this requires the setting-up of distinct and specific processes, which of course adds to the complexity of its operations. Secondly, the incubator is not usually a simple and neutral venue where entrepreneurs and economic actors interact freely and autonomously. Interactions between start-ups and other economic actors are inherently laden with uncertainty and risk, not to mention significant information asymmetries that may be present on either side, thus leading to moral hazard. In order to operate as an effective economic exchange, the incubator must therefore build a significant *reputational capital* (Hoppe and Ozdenoren, 2005) by running a careful screening process of the actors it involves and by supporting their interactions so as to minimize the risk of disruption and litigation. For the parties involved, interaction occurring within the boundaries of and overseen by the incubator provides a significant guarantee. These parties know that the incubator has a strong incentive to avoid problems that could damage its reputation and the value of its matchmaking activity. This reputational aspect is probably more relevant in start-up ecosystems that are not fully developed, and where the 'rules of the game' and the ethos of what it means to run – or cooperate with – a start-up are not yet widely known. Conversely, an incubator operating in an ecosystem with many experienced entrepreneurs and economic actors can probably limit itself to connecting them, and then leaving ample freedom for the parties involved to decide whether and how to proceed. In any case, the ability to work on individual connections and building reputational capital is probably the reason why incubators continue to exist and have not been supplanted by internet exchanges such as AngelList.

Bringing the two previous roles together, the incubator may also be viewed as a mini *business cluster* exhibiting *economies of agglomeration* (Porter, 1996). This perspective is interesting, because of the potential value of inter-firm relationships between start-ups. These relationships may occur at a formal level, when complementary start-ups cooperate with one another to develop complex products and services, or when competing start-ups cooperate or engage in small-scale merger and acquisition (M&A) activities to achieve economies of scale and accelerate their evolution. Inter-firm relationships may also be informal and based on knowledge, experience and network sharing between individuals in an entrepreneurial community (Feld, 2011) including both current start-ups and alumni. It is uncertain whether this latter type of interaction should be considered as a complement or as a substitute for the formal consultancy and matchmaking activities carried out by incubator coaches and mentors. Should it be a substitute, it may lead incubators to progressively shrink their activity as the entrepreneurial community they have nurtured grows.

If one focuses on *university* incubators, a further role as a *bridging institution* for technology transfer emerges (Kremic, 2003). A significant part of a university incubator's activity is in fact associated with creating opportunities to exploit research results by supporting the launch of start-ups that may license such results and bring them to market. In this case, they follow a *technology-push* process (in other terms 'given the solution, what is an addressable problem and market?'). Moreover, university incubators often work on *demand-pull* innovation processes by supporting start-ups that answer a clear market need to tap into relevant technical competencies at the university. When dealing with the former type of process, the incubator will have to manage a potential conflict of interest between its primary duty towards entrepreneurs, which calls for maximizing the value of the start-up, and its institutional affiliation to the university, which instead might call for maximizing the returns accruing to the same institution. In order to avoid getting stuck in embarrassing positions and undermining its reputation, the incubator must ensure that regulations governing academic entrepreneurs and their relationship to the parent institution are clearly spelled out in advance.

In any case, acting as a successful bridging institution is likely to require an affiliation with a strong university, able to produce a variety of high-quality and market-relevant research and competencies. On the other hand, the bridging institution must have sound connections with the associated industries. Moreover, given the important differences between the technology-push and demand-pull processes as outlined above, the incubator must be able and flexible enough to manage both.

Table 1.1 attempts to summarize the previous discussion, pointing out the four economic roles, their possible economic drivers, the resulting key success factors and the related implications for incubator managers.

Based on this representation of the economic foundations of a university incubator, and taking into account the disclaimer that it represents a set of hypotheses, it is possible to draw some tentative conclusions

Table 1.1 A summary of the possible economic determinants of academic incubation

Economic role	Economic drivers	Main actors and key success factors	Strategic implications
Consultancy company	Learning economies (generic with respect to start-ups) Learning economies (industry-specific) Scale economies (allowing a hierarchy of consultants)	Coaches and mentors (and their ability as consultants)	Deciding whether to pursue a generalist or specialist strategy Deciding the degree to which coaches and mentors can play an active part in start-ups' decision-making processes
Two-sided market	Economies of scale and scope on the multiple markets involved Reputational capital and ability to support transactions with high uncertainty and information asymmetries	Size, variety and quality of the pool of start-ups Size, variety and quality of the external network Coaches and mentors (and their ability as matchmakers)	Deciding the effort to be spent on building the different markets and the directions to take
Industry cluster	Economies of agglomeration (scale and scope)	Size, variety and quality of the pool of former and current start-ups Richness, experience and connectedness of the entrepreneurial community	Deciding the effort to be spent on creating inter-firm connections among start-ups and alumni Monitoring inter-firm collaboration and understanding its role (substitute or complement) with respect to the activity of coaches and mentors

Table 1.1 (continued)

Economic role	Economic drivers	Main actors and key success factors	Strategic implications
Bridging institution	Economies of scale and scope Learning economies	Quality, size and variety of academic competencies at the affiliated university Clarity in the regulations governing academic entrepreneurship Coaches and mentors (ability to operate both according to technology-push and demand-pull)	Deciding the effort to be spent on technology-push and demand-pull activities Building the network with academic labs and departments

that – again – must only be considered as a stimulus for debate and as possible directions for further research.

1.3.2 Ownership and Governance

Just like any other economic entity, the profile and the strategic direction taken by a university incubator will be heavily dependent on its ownership structure. An incubation program run by an administrative office of a university, a non-profit consortium such as I3P, and a for-profit incubator jointly owned by a university and an investment fund, will be subject to different charters and missions, will follow different strategies and – of course – will achieve different results. As a consequence, the connection between ownership structure and the performance of an incubation program should not be overlooked, and must be taken into account both when defining the former and when assessing the latter.

Closely connected to the issue of ownership is that of governance. As shown in the previous subsection, incubators operate a complex business model that is not easy to frame in terms of modes of value creation, assessment of the costs incurred, evaluation of the direct impact and the externalities generated. Moreover, incubators often operate at the intersection between for-profit and non-profit sectors. Therefore, the governance of an incubator must define clear objectives and metrics for evaluating the program, taking into account the long time constants over which many of these indicators will develop. The revenue model of an incubator must also be coherent with the expected outcomes, so that its operations can run in a clear and unambiguous way. In the case of I3P, having revenues linked to start-up creation has indeed led to a clear and strong incentive to generate a strong flow of potential start-ups and a significant number of new ventures. However – as discussed – this may not have generated a sufficiently strong incentive mechanism to work on the growth of these companies, which is something that could be corrected by creating a parallel and for-profit acceleration program.

Finally, and especially for incubators that are supported by public funds, a constant monitoring of the market conditions must be carried out. Specifically, it is appropriate to check whether a market failure justifying public intervention is still present or whether – perhaps as a result of the work carried out by the same incubator – the start-up ecosystem has become sufficiently mature to operate independently. Failing to do so opens the risk of *crowding out* private players who might offer the same supporting role to start-ups (Thierstein and Willhelm, 2001).

1.3.3 Vertical and Horizontal Integration Choices

The dual nature of the incubator as a consultancy company and a two-sided market also involves making key strategic choices with respect to vertical integration. For example, legal advice to start-ups can be provided by an in-house lawyer or by a partner law firm. In the former case, there will be a tighter control over the resource (provided incubator managers are able to hire and manage legal experts) and a higher fixed cost. In the latter case, the incubator will have lower costs, less control over the resource, but may partner with multiple service providers, thus enriching the choice available to start-ups and lowering the risk associated with non-performance by the resource.

Furthermore, a higher degree of vertical integration (i.e. in-house experts) will tend to concentrate valuable and specific expertise within the incubator. Conversely, a lower degree of vertical integration (i.e. partnering with outside experts) will spill these competencies into the local start-up ecosystem (i.e. the partner lawyer will learn how to deal with start-ups beyond those that operate in the incubation program). Therefore, vertical integration choices will be made based on costs and competencies, but may also take into account the degree of strategic control the incubator wants to have over key resources in the local start-up ecosystem.

This aspect is particularly critical when dealing with the core consultancy that is provided by incubators, namely start-up strategy. At its heart, this is about choosing the right mix of coaches and mentors. As mentioned in the earlier footnote, the former are usually employed and paid by the incubator, while the latter are independent partners who work *pro bono* or are paid directly by the start-ups (usually with success fees). When starting an incubation program in a strong start-up ecosystem, it is likely that good mentors will abound, and coaches may not be needed at all. However, when an incubator starts operating in a weaker ecosystem, mentors might not exist (and aspiring ones might not be credible enough), thus making it mandatory to train and develop a team of coaches. As time goes by – and as the ecosystem evolves – it is likely that the optimal mix between coaches and mentors will change. However, it might not be easy to manage this shift, due to organizational inertia and the fear coaches may have of losing their jobs. The choice made by I3P to rely heavily on coaches while contracting them as freelancers could be seen as a way to keep the middle path in this strategic choice.

Horizontal integration is choosing to specialize by industry or by phase of the start-up development lifecycle. As already mentioned, the right balance must be found between the benefits accruing from economies of scale and learning both across the board and within each possible specialization.

1.3.4 Organizational Aspects

Regardless of the strategic choices made, it should be recognized that an incubator is a firm, albeit a peculiar one. Viewing it as a development agency or an academic institution would probably undermine the credibility and capability of the incubator's staff to support start-ups effectively. In order to support entrepreneurs, one must share the same culture and be subject to similar dynamics. This is of course difficult if one operates within an entity with a different culture, processes and set of incentives.

That said, and based on the four economic roles discussed above, a number of operational decisions must be made. The most important decisions are probably associated with the management of the incubator's human resources, including recruitment, incentive schemes, professional development and organization. After all, an incubator is a 'people business', and the quality of its human resources and the way they are organized will determine its capability to perform effectively.

The importance of creating and sustaining external relations at senior

Table 1.2 Required capabilities for incubator coaches

Role	Typical competencies and skills
Consultancy	• Analytical skills, for understanding markets and industries • Competencies in innovation management, for framing and proposing strategy • Competencies in finance, for interfacing with investors and banks • General competencies in administrative, financial, legal and IP affairs (gatekeeping with professional services providers) • Communication skills, both concerning their personal capability to present and for tutoring entrepreneurs • People management skills, for coordinating the work of junior consultants and analysts • People management skills, for coaching entrepreneurs (understanding their personal shortcomings and defining ways to overcome them)
Network creation	• Personal credibility and strong understanding of business dynamics, in order to create links with existing firms • Personal credibility and strong understanding of the technological state-of-the-art, in order to create links with academic staff
Deal-making	• Capability to screen people and avoid engaging in impossible deal-making processes • Analytical skills for supporting negotiations • Communication and personal skills for supporting negotiations

level is a strong requirement especially of the program's top management (CEO and/or managing director). At the same time, coaches (and mentors) must possess a significant variety of skills and competencies given the manifold nature of their work, as summarized in Table 1.2.

The skill set required for coaches is very broad and not easily found on the market. This suggests that an incubator should train these key personnel 'in-house' rather than attempt their external recruitment. Recruiting from other incubators is of course possible, but carries the risk that experience gained in a different start-up ecosystem might not be applicable in the new location. Another possibility is looking for people with a similar skill set, for instance in the management consultancy industry. However, caution is advised because there is a big difference between working with start-ups and established companies.

Finally, the competencies and skills listed above apply with different weights and nuances along the lifecycle of a start-up. An incubator might therefore decide to split the incubation process into phases and have *different* and specialized start-up coaches looking after each phase. The trade-off here is between the benefits of specialization or the value of having a single, well-recognized coach (who entrepreneurs get to know and trust) taking the start-up all the way through the process from beginning to end.

1.4 CONCLUSIONS

This chapter has attempted to discuss the nature of academic incubation based on the experience of a single and well-established program, i.e. the I3P incubator of Politecnico di Torino. Based on this experience, the author has attempted to build an economic analysis of what constitutes an academic incubation program by highlighting four roles, i.e. that of a consultancy company, a two-sided market, an industry cluster and a bridging institution. This analysis constitutes a basic theoretical and unvalidated discussion. However, this could lead to further empirical research, in order to gain a deeper understanding of the economic foundations of academic incubation.

Then, in order to provide some preliminary practical advice, this analysis has been used to highlight three main areas where incubator managers must take key decisions: ownership and governance, vertical and horizontal integration, and organizational policies.

NOTE

1. Here, and for the remainder of the chapter, we will use the term *coach* to describe incubator personnel acting as consultants to start-ups. In the incubation industry, this role may also be referred to as *tutor*, *consultant* or *expert*. Coaches are different from *mentors*, the latter being independent experts with an entrepreneurial or managerial background, who share their experience and their networks with start-ups either *pro bono* or via a contractual link with the start-ups. However, the incubator does not usually pay mentors.

REFERENCES

Cantamessa, M. (2015), 'Verso la Entrepreneurial University del XXI Secolo', *Opening lecture of the 2015–2016 Academic Year*, Politecnico di Torino (in Italian).

Chesbrough, H.W. (2006), *Open innovation: The new imperative for creating and profiting from technology* (Boston, MA: Harvard Business School Publishing).

Erkko, A. and H. Yli-Renko (1998), 'New, technology-based firms as agents of technological rejuvenation', *Entrepreneurship and Regional Development*, **10**, 1.

Etzkowitz, H. and L. Leydesdorff (2000), 'The dynamics of innovation: from National Systems and "Mode 2" to a Triple Helix of university–industry–government relations', *Research Policy*, **29** (2), 109–23.

Feld, B. (2011), *Startup communities: Building an entrepreneurial ecosystem in your city* (Hoboken, New Jersey: Wiley).

Hekkert, M.P., R.A.A. Suurs, S.O. Negro and S. Kuhlmann (2007), 'Functions of innovation systems: A new approach for analysing technological change', *Technological Forecasting and Social Change*, **74**, 4.

Hoppe, H.C. and E. Ozdenoren (2005), 'Intermediation in innovation', *International Journal of Industrial Organization*, **23** (5), 483–503.

Kremic, T. (2003), 'Technology transfer: a contextual approach', *The Journal of Technology Transfer*, **28** (2), 149–58.

Moretti, E. (2012), *The new geography of jobs* (New York: Houghton Mifflin Harcourt).

OECD-EC (2012), 'A guiding framework for entrepreneurial universities', http://www.oecd.org/site/cfecpr/EC-OECD%20Entrepreneurial%20Universities%20Framework.pdf (accessed 30 April 2016).

Peters, L., M. Rice and M. Sundararajan (2004), 'The role of incubators in the entrepreneurial process', *The Journal of Technology Transfer*, **29**, 1.

Porter, M.E. (1996), 'Competitive advantage, agglomeration economies, and regional policy', *International Regional Science Review*, **19** (1–2), 85–90.

Ries, E. (2011), *The lean startup: How today's entrepreneurs use continuous innovation to create radically successful businesses* (New York: Random House).

Rochet, J.C. and J. Tirole (2006), 'Two-sided markets: A progress report', *The RAND Journal of Economics*, **37** (3), 645–67.

Simon, M., S.M. Houghton and K. Aquino (2000), 'Cognitive biases, risk perception, and venture formation: How individuals decide to start companies', *Journal of Business Venturing*, **15** (2), 113–34.

Thierstein, A. and B. Willhelm (2001), 'Incubator, technology, and innovation centres in Switzerland: features and policy implications', *Entrepreneurship & Regional Development*, **13** (4), 315–31.

Zomer, A. and P. Benneworth (2011), 'The rise of the university's third mission', in J. Enders, H.F. de Boer and D.F. Westerheijden (eds), *Reform of Higher Education in Europe* (Rotterdam: Sense Publishers).

2. Strategies for designing new venture units in complex contexts

Elco van Burg, Isabelle M.M.J. Reymen, A. Georges L. Romme and Victor A. Gilsing

2.1 INTRODUCTION

Large, mature organizations are often capable of exploiting existing products efficiently, but are typically less effective in being innovative. Financial systems and bureaucratic procedures adopted to control processes in large organizations tend to be hostile towards innovative ideas, proposals and initiatives. One of the solutions to this problem is to structurally separate exploitation tasks and innovative exploration activities, the latter, for example, in a new venture unit. On the other hand, such a structurally separate unit still needs to have some degree of integration with the parent organization, which forms the lifeline for new ventures in terms of resources and reputation. As such, the new venture unit acts as an 'incubation' semi-structure that mediates organizational rigidities and supports organizational renewal by means of entrepreneurship. Previous studies have provided detailed assessments of the layout of such a new venture unit and its simultaneous integration with and separation from the host organization (e.g. Jansen et al. 2009). However, *how* these units are established in the first place has largely remained unaddressed.

In this respect, our understanding of the process of designing such units is still in its infancy, and studies considering how designers use knowledge to deal with the complex contexts of this design process are rare. Here, this study contributes to the innovation and corporate and academic entrepreneurship literature by studying the interaction between the design processes of new venture units and diverse complex design contexts. The way designers use and process knowledge can be conceptualized in terms of three design strategies (Gavetti et al. 2008): off-line reasoning and planning, feedback-driven learning and associative reasoning. Research on designing new venture units implies that in many organizations this design process is especially driven by experimentation (i.e. feedback-driven learning) or by copying designs (i.e. associative reasoning) from other organizations (Hill and Birkinshaw 2008). An important question then is how specific contexts enable or hamper particular design strategies.

This chapter focuses on the processes of designing new venture units in complex contexts such as universities. The empirical part of the study draws on three case studies of the creation of three new venture units at universities. In these three cases, we study the design strategies used to construct such a separate yet bridging unit in a university setting.

This chapter is structured as follows. First, we explore how designers use design strategies to deal with specific contexts. Next, we discuss important challenges and contextual factors related to organizing new venture units. Subsequently, the empirical setting is described. The 'Method' section discusses the case selection, data collection and data analysis. We then describe and interpret the design processes at the three sites. Finally, our main findings are discussed.

2.2 THEORY DEVELOPMENT

2.2.1 Organization Design and Design Strategies

Recently, a more deliberate focus on design *processes* has resulted in methodologies that serve to develop design principles from practitioners' knowledge and from research synthesis (e.g. Van Burg et al. 2008). This line of research recognizes contextual factors as being critical for design processes and the performance of designers in terms of efficiency and effectiveness. Although several authors emphasize the contextuality of organizational knowledge, research on organization design (processes) has not yet addressed how contextual knowledge influences design processes.

The design context can be characterized as a complex organizational system, which according to Ethiraj and Levinthal (2004: 407) consists of 'a large number of elements that interact in a non-simple way'. For Herbert Simon (1962), the designability of these systems is primarily influenced by their degree of hierarchy and their near-decomposability (or loose coupling). *Hierarchy* implies that the system is composed of subsystems and also involves subordination in an authority relationship. Therefore, in organizations with a high degree of hierarchy, design decisions at higher hierarchical levels can be enforced at lower levels. Conversely, conflicts between subsystems can be resolved by decisions at higher hierarchical levels. Moreover, hierarchy also involves the notion of nested hierarchies. These nested hierarchies enable local search and trial-and-error designing while the total system remains stable.

Near-decomposability, also known as loose coupling, refers to the clustering of interactions between agents within each of these subsystems instead of dispersed interactions between all agents in the system. In other words, in a nearly completely decomposable system, the rates of interaction within each subsystem are much higher than the rates of interaction between the subsystems. This bundling of interactions in subsystems reduces the cognitive demands on the designer (Ethiraj and Levinthal 2004). Another advantage of near-decomposability is that it enables localized adaptation within each of the subsystems.

In sum, hierarchy and near-decomposability are two important characteristics of the organizational context to be considered in design processes. These characteristics are, however, not identical with the designer's representation of them. 'How complex or simple a structure is depends critically upon the way in which we describe it' (Simon 1962: 481). The design efforts to deal with this complex context are guided by three 'modes of cognition', or design strategies, employed by the designers (Gavetti et al. 2008): off-line reasoning and planning, feedback-driven learning and associative reasoning.

First, *off-line reasoning and planning* is the cognitive assessment of the design problem by the designer and the search for potential design solutions in a specific context without engaging in interactions with this and other situations (Gavetti et al. 2008). This assessment can be supported by computational approaches and systems. As a design activity, off-line reasoning and planning is fairly straightforward, does not engage in actual interaction with the context and involves reasoning based on existing knowledge available to the designer. In this design strategy, the consistency and processual logic of designing is especially emphasized. In this respect, Simon (1996) outlined how data, planning and forecasting may enable high-performing designs, adapted to the specificity of the design situation. But he also identified the main limitation of this computational approach, namely that rationality is bounded. This limitation especially applies to social systems (e.g. organizations), for which 'the complexity of the environment is immensely greater than the computational powers' (Simon 1996: 166).

Second, by *feedback-driven learning* the designer generates design knowledge by actually engaging in design actions and by processing positive or negative outcomes of these actions, leading to adapted or reinforced strategies (Gavetti et al. 2008). By actually creating the design artifact in interaction with the context, the designer tests whether understanding of the context is correct and to what extent the design should be adapted to fit in the context to achieve a desired situation (Romme and Endenburg 2006). In this respect, Donald Schön's influential work on 'reflective practice' argues that the context (e.g. infrastructure, clients, users) 'talks' to the designer, who subsequently has to reflect on it and use these reflections in order to create a good design (Schön 1984, 1987: 44). By means of this conversation the designer creates understanding of and gives meaning to the context.

Third, *associative reasoning* is the process in which the designer generates knowledge by comparing the situation at hand with other situations, in terms of analogy, case-based reasoning or imitation. According to many scholars, the implicit or explicit use of analogies is central to human thinking and is typical for designers (e.g. Lawson 2006; Schön 1984). By framing the context, the designer can imagine meaningful design actions and understand some of the essential characteristics of the situation. Gavetti et al. (2005) model the situation in which a designer draws on analogies to make design decisions. Whereas there are major difficulties in making the correct analogies – in terms of getting the right source and getting the transfer context right – Gavetti et al. (2005) find that any analogy, even irrelevant ones, provides better results than making only local decisions without analogies outside the existing framework. The story recounted and interpreted by Weick (1990) about the troops that found their way out of the Alps with a map of the Pyrenees illustrates this (Gavetti et al. 2005). Moreover, good analogies produce better results than bad ones. Here, the designer's experience moderates the transformation of design knowledge of context A into design knowledge that is valid in context B. In this respect, the designer's experience can be imagined as a repertoire of analogies s/he can draw on. The larger this repertoire is, the more likely the designer will select an appropriate analogy for transforming knowledge from one context into another.

2.2.2 Organization Design for New Venture Creation

This study explores design activities and interactions in different contexts to examine how designers deal with designing new venture units in large, complex organizations. To introduce this particular type of design, we briefly review the literature on new venture creation in complex contexts. We focus especially on new venture creation from universities and discuss relevant contextual factors.

A large number of studies in the innovation literature recommend designing separate divisions or units to support the creation of new businesses alongside mainstream operations and as an extension of R&D activities (e.g. Ambos et al. 2008). Such a structurally autonomous unit helps to decouple rigidities in the parent organization from the actions and perceptions in the new venture. Researchers have been less concerned with design processes at the level of the new venture units (Hill and Birkinshaw 2008). In this respect, Hill and Birkinshaw (2008) observe that in many organizations this design process is driven by experimentation or by copying designs somewhat unreflectively from other organizations, even to inappropriate settings. Similarly, designing new venture support units and policies at universities creates tensions with existing practices. These tensions include commercial goals versus research goals, the ownership of intellectual property and the allocation of revenues. The design of these units and the performance of design processes in terms of efficiency (i.e. effort and time needed) and effectiveness (i.e. arriving at the intended results as well as the viability of these results) is influenced by multiple contextual levels (Bekkers et al. 2006; Hill and Birkinshaw 2008). In the empirical study in this chapter, we focus on two levels: the national and regional level reflecting the policy context; and the university level reflecting the organizational context.

At national level, new venture creation by universities is particularly influenced by intellectual property (IP) laws, such as the Bayh–Dole Act in the US. At both national and regional levels, several public policies influence new venture creation from universities, such as governmental subsidies, the business and entrepreneurial climate, educational policies and access to the capital market (Bekkers et al. 2006).

At university level, the size of the organization increases the complexity and thus tends to decrease the ability to redesign the whole organization. However, size may also invoke more change and reorganization. The formal organization of a university can be considered as a professional bureaucracy in which the hierarchy is anchored in the consultation processes and procedures with a high degree of decentralization and loose coupling. Besides that, a high degree of politics complicates decision-making, whereas organizational hierarchy may serve to reduce politics. Furthermore, the focus, eminence and culture of the incumbent university influence new venture creation. This contextual factor, however, is very hard to influence and design in the short term. More instrumental in this respect are the policies and procedures, including incentives, for new ventures, and the new venturing support infrastructure (e.g. advisors, incubators) that a university has. With regard to policies and infrastructure directly targeted at new venture creation, the experience in new venture creation and the organization of the support process affect the quantity and quality of the new ventures created.

Although several studies have been conducted regarding the establishment of support structures, the context-specificity of designing new venture units has not been explicitly addressed. Our study therefore theorizes about the use of different design strategies in specific contexts as well as how these strategies relate to the performance of the design process.

2.3 METHOD

To examine how complex contexts interact with the use and performance of design strategies, we performed a descriptive study of the (re)design of three European universities of technology with regard to facilitating and stimulating the creation of new ventures (also called 'university spin-offs'). The three cases selected for this study are: Barcelona Tech (UPC) in Spain, and Eindhoven University of Technology (TU/e) and Wageningen University (WU) in the Netherlands. These three universities share a strong focus on science and technology, which provides fertile ground for new venture creation. Each of the three universities performs a critical role in its region, both in terms of education and economic development. With regard to economic development, all three universities seek to increase knowledge and technology transfer from their university to the market, to contribute to societal and economic development.

At all three universities, data were collected from multiple sources and backgrounds to get a rich understanding of the design processes. Interviews were conducted with the initiators of the programs, different support professionals, university professors and deans. Moreover, policy-makers and support professionals from organizations and government bodies outside the university were interviewed. Furthermore, at each university, academic entrepreneurs were interviewed. These entrepreneurs were chosen to be representative of the main characteristics of the local population of academic entrepreneurs, with regard to experience, age, disciplines and departments, period of starting the company, as well as successes and failures. In total, 64 hours of interviews with 66 people were recorded and fully transcribed. In addition, different types of documentation were collected, including annual reports, strategy documents of new venture support units, business plans of the new ventures, and other studies by the universities.

In the analysis of the cases, the first step was to construct a comprehensive overview of the design and development of the university and the new venture support organizations within the university (see Table 2.1). The second step was to code, within each of the cases, the main characteristics and the main design actions. This resulted in a design narrative for each case. Third, design actions were related to the three design strategies proposed by Gavetti et al. (2008). This served to inductively 'measure' the 'amount' each strategy was used: the more design actions in a case were associated with one of the strategies, the more important that strategy was in that specific context (Table 2.2). Subsequently, for each of the design strategies, we searched for case evidence indicating the relationship between the employment of these strategies and the performance of the

Table 2.1 *Main characteristics of the universities*

Characteristics	UPC	TU/e	WU
National and regional			
Intellectual property rights	University owns the IP of inventions by employees. Free usage of national patents by a university.	University owns the IP of inventions by employees, but the university compensates the inventor for developing the patent.	
New venture and higher education policy	Since 2001, structural funding for new venture support. Until 2006 the inventor's equity was limited to 10%, now staff members can have more than 10% if the university also gets a stake.	Since the 1990s, several funding schemes for new venture support exist; these funding schemes have become more structured since 2002. Academic staff allowed to have equity in companies.	
Access to capital	Many subsidy schemes are established, especially since 2001. The importance of venture capital and informal investors is increasing, but still moderate regarding investing in new ventures.	Many (national) subsidy-schemes have been established, especially during the last 10 years. The importance of venture capital and informal investors is increasing, but still moderate with regard to investments in new ventures. These investments are also closely related to particular sectors (e.g. biotech), and less to national or regional characteristics.	
Regional clusters	Strong high-tech cluster, especially with regard to biotech and information technology.	Strong high-tech cluster, mainly due to the proximity of several large multinational firms in electronics and chemistry.	Strong life sciences cluster. In addition, an incubator for new ventures was established in 2001.
Interactions with policymakers	Very high	High	Low

Table 2.1 (continued)

Characteristics	UPC	TU/e	WU
University			
Near-decomposability	Moderate	Moderate	High
Hierarchy	High	Moderate	High
Size	2006: around 35,000 students and 4,000 employees.	2006: around 7,000 students and 5,000 employees.	2006: around 9,000 students and 6,000 employees.
Organization	41 faculties and 19 other departments (among which the Programa Innova); election of the vice-chancellor by employees of the university.	9 faculties and 12 other departments (among which Innovation Lab). Board of Directors appoints the vice-chancellor, who is responsible for education and research, as well as the president who is responsible for general management of the university.	8 faculties and 21 other departments (among which Wageningen Business Generator (WBG)). Board of Directors appoints the vice-chancellor, who is responsible for education and research, and is also responsible for general management of the university.
History	Founded in the 1970s with a focus on engineering and architecture. Later UPC broadened its scope towards others fields and locations, partly as a result of a number of mergers. Has a long history of contract research with industry.	Founded in the 1950s with a focus on electronics and industrial engineering. Since the beginning, TU/e has had strong ties to industry.	Founded in the 1880s, with a focus on education for agriculture. Merged in 1997 with a research institute and in 2004 with a professional education institute. Has a long history of collaboration with industry and government.

Intellectual eminence of the university	THES* ranking 2007: none	THES ranking 2007: 100–150.	THES ranking 2007: 100–150.
Overview of new venture policies and support infrastructure	Since 1998, UPC is supporting new ventures, especially targeted at training and culture change. The university rarely invests in companies. Limited facilities and incubation space.	Since 1998, TU/e has an explicit new venture policy by establishing an incubator and making substantial equity investments in companies (up to 100%). Since 2004, there is more support for new ventures. Patents are usually sold to the companies for a certain equity and royalty share.	Since 1999, WU has defined policies to stimulate new ventures. Since 2004, WU has a new venture policy targeted at investments in high-potential companies, and provides considerable support. In turn, WU acquires a substantial equity share. Patents are usually sold for an equity share.
Financial	2006: €26 M revenues from industry collaboration, revenues from patents €0.1 M (total turnover €269 M).	2006: €45 M revenues from industry collaboration; revenues from patents and new ventures not public (total turnover €265 M).	2006: €169 M revenues from industry collaboration, €6 M from patents and new ventures (total turnover €590 M).

* Times Higher Education World University Rankings.

design processes. The performance of the design process is operationalized as (1) efficiency of the design process in terms of the efforts and time needed and (2) effectiveness of the design process in terms of arriving at the intended results (i.e. a unit supporting new venture creation) and the viability of the design results (i.e. survival of the unit). Fourth, the three cases were compared along the different identified context characteristics and design strategies, using tabular representations.

2.4 RESULTS

This section describes the results of the case studies. An overview of design and development of policies and activities with regard to new venturing in the three universities is provided in Table 2.1. More comprehensive case narratives are available in Van Burg (2010). The main components of the design of a new venture unit at the three universities are discussed in terms of the three design strategies. Next, the contribution of the design strategies employed to the performance of the design process is discussed.

2.4.1 Using Three Design Strategies

We discuss the three design strategies by considering how their usage varies over the three context characteristics. Table 2.2 provides an overview of important design actions in the three cases and the related design strategies.

Off-line reasoning and planning by means of careful analysis of available data was observed in all cases, especially relating to funding processes. Government agencies, university boards, development programs and other financers requested extensive documents to assess the plans (e.g. for funding new units). These plans typically included descriptions of the goals, anticipated support activities, the relationship with other initiatives, existing support and expertise, and, most importantly: a budget proposal. These documents represented the design as a planned exercise set up to produce the results expected by the funding agencies. As a result, such plans often represented a 'political' reality in which numbers, reasoning and action sequences were primarily a means to get the funding, only to a lesser extent reflecting the real activities. An entrepreneurship professor at TU/e reflected on this 'reported' reality: 'In the past the support was just "window dressing". . . . The government says: universities have their societal responsibility, and we will provide the money. Then, as a university, we started to make a lot of smoke, but the fire was hardly visible, at least not within the university.'

Regarding the context, the degree to which agent-designers adopted

Table 2.2 Design actions and associated strategies

Case	Year	Important design actions	Off-line reasoning and planning	Feedback-driven learning	Associative reasoning
UPC	1998	Creating Programa Innova	x		x
	1999	Adapting Programa Innova		x	
	2000	Expanding, more structural support		x	
	2001	Access to seed capital			x
	2002	Adapting seed capital program		x	
	2004	Network development			x
	2004	Adapting seed capital program		x	
	2005	Hiring IP advisor		x	x
	2006	Restructuring support around ventures		x	
	2007	Structuring into one network		x	
	2007	Creating program targeted at students			x
	2007	Adapting seed capital program		x	
	2007	Converging programs		x	
WU	1998	Developing business support office	x		x
	2001	Founding (external) incubator	x		x
	2002	Redesigning strategic and financial aspects		x	x
	2004	Founding WBG	x		x
	2005	Creating network organization			x
	2008	Restructuring WBG		x	x
TU/e	1998	Founding incubation center	x		x
	1998	Founding university holding structure			x
	2002	Developing new strategy, focused on spin-offs (more support and services)	x	x	x
	2004	Redesign of support organization, with one orchestrating unit (Innovation Lab)	x	x	x
	2005	Creating network organization	x	x	x
	2006	Adjusting policies: e.g. equity as payments		x	x
	2008	Embedding support in faculties	x	x	

off-line reasoning and planning depended primarily upon the history and phase of the support, the hierarchy of the organization and the interaction with especially national or regional policymakers. The history of new venture support at the incumbent university determines what is already known and accepted within the organization. This acceptance influences the degree to which redesign efforts have to be presented as careful reasoning and planning in order to convince organization members. Within UPC, these off-line reasoning and planning activities were rare and limited at the start of the program in 1998. At TU/e and WU, more recent explicit redesigns implied that funding agencies required extensive motivation and planning, presented as careful planned redesign. The degree of formal hierarchy and the place of the designers in this hierarchy affected the extent to which the (re)design was justified by documents, careful analyses, planning and the presentation of results. Whereas the primary knowledge sources for the redesign plans were largely provided by associative reasoning or the emergent results of feedback-driven learning (for instance, the delivery of successful ventures), the plans were ultimately written and presented as carefully planned designs created by off-line reasoning. These documents also served to shape design efforts and increase commitment at various levels within the university. The same occurred at the national level. As such, off-line reasoning and planning provided an important strategy to guide the interaction with policymakers. The director of TU/e's Innovation Lab stated:

> In about five years the University of Twente produced about 480 new ventures. Rotterdam produced 11, Leiden 2. The rest of the universities were not on that list. The Secretary of State concluded they did not produce new ventures. . . . I said: well, I believe we at least have a professor of entrepreneurship over here. . . . But, we simply did not fill out these charts.

So, in the interaction with policymakers and decision-makers, official planning documents and preliminary results were important to convince them and gain their approval. To achieve this approval, the director of TU/e's Innovation Lab subsequently made sure that he presented numbers and detailed plans of the redesign activities, although they were often primarily written to convince policymakers and decision-makers rather than to shape the actual design activities.

Feedback-driven learning by frequent interactions with academics and academic entrepreneurs appeared to facilitate learning with regard to what does (not) work within a certain setting. At UPC, for example, the new venture unit discovered through interactions that it was not feasible to change the mindset and priorities of academics or to allow all of them to engage in entrepreneurial activities. The entrepreneurship support programs were therefore made available from a certain distance, to make sure that they did not interfere with education and research processes and did not need the commitment of all faculty members. At TU/e, an important adaptation was to bring technology transfer, contracting with industry, intellectual property support and entrepreneurship support under one umbrella to align policies and create an integrated approach. Previously, these activities were distributed across different departments, which was widely perceived as problematic and confusing. Combining these activities under one aegis made them easier to access and integrate. Another example from the TU/e case involves equity participation. TU/e's initial demand for 100% equity in new ventures resulted in some major failures – the CEOs of these companies had no exposure to ownership incentives or risks. Over the years, TU/e therefore moved to minority equity participations. Furthermore, TU/e also began to treat equity as a form of payment by the entrepreneur for the invention and support. A start-up support officer explained:

> At the department of chemistry, we had this situation where we said: if you want to stay here, these are the costs. But the venture could not afford that. So we [the new venture unit] said: we will pay the bills to the department, if we get a little tuft of your equity in return.

This instance illustrates how active learning occurs in response to feedback, adapting practice as insight emerges.

In all three cases, feedback-driven learning appears to be important, but the specific organizational context affects the frequency of feedback-driven learning and the organizational level (e.g. board, department and faculty) at which useful feedback can be obtained. First, the degree of near-decomposability apparently affects the opportunities for frequent interactions. For example, WU departments are relatively autonomous and the directors of these departments control the implementation of Wageningen Business Generator (WBG) support. Some departments did not commit to the new venture support activities and policies, and therefore interaction with these departments was limited. Because interests were too diverged, these interactions did not result in many adaptations at the central WBG new venture unit. Instead, they resulted in new strategies being developed at departmental level, thus causing divergent interactions. In effect, department directors and faculty learned how to evade the consequences of the central organizational support policies. The director of WBG stated:

> This [the structure of WU] creates enormous complexity! Because people have many escape routes. . . . The director of department X, who is also responsible for the portfolio of commercialization, used decentralized approaches and tried

> everything to obstruct [WBG] corporate activities. We always ran into trouble. If we had a difference of opinion with an inventor and we said: 'We are not going to invest, because this case is not good enough', then the inventor would go public with this and get lots of media attention. . . . New ventures were founded outside WBG and the university board accepted that.

At both TU/e and UPC, however, the new venture support unit became more embedded in departments and less perceived as just part of the central organization. Thus, frequent interactions with the individual groups and researchers within departments served to develop tailor-made programs. One academic entrepreneur at TU/e illustrated this interaction and the flexibility: 'No question is too crazy, and often the answer is: okay.'

Second, organizational hierarchy influences especially the level (board, department directors, research groups or individual faculty) at which feedback-driven learning occurs. At both WU and UPC, formal hierarchy was important; at UPC the formal hierarchy was reinforced by political power. In both cases, decisions were typically taken top-down. As a result, learning occurred at positions higher in the hierarchy, especially in interaction with external organizations and departmental directors and deans. Less interaction was observed with academics and other workers at the operational level of the university. For instance, one of the former support officers at UPC observed:

> Any program that involved the faculty members on a regular basis in the process never appeared to work. We tried this, we tried that, and in the end, we concluded that the only ally we had in accomplishing the things we wanted was [the initiator and the current director of the program]. Because he is an opinion-leader in the university.

In this respect, Programa Innova at UPC learned about faculty disinterest from resources available to carry out support tasks, and subsequently focused on interactions at higher levels of the university organization (board, directors and deans) and interactions with public policymakers. As a result, the design of the support program became more targeted at developing different funding models in interaction with representatives of other governmental agencies, because these relationships were close and well developed. At WU, the situation was similar because of limited integration in the departments, so WBG focused on executing the support design without much adaptation. However, because of the lack of embeddedness at lower levels of the organization (i.e. departments, research groups and faculty), design activities appeared to be vulnerable to decreasing support of the board, which could cause rapid changes in the new venture support policies and activities. The major reorganization that took place in the WU case illustrates this. The director of WBG, responsible for designing and executing new venture support at WU, stated:

> Some [of the department directors] supported me, but others were somewhat more opportunistic. They saw a strategy emerging: 'the boss [the vice-chancellor] says something, he wants this'. That is the reality of a political situation. . . . This was basically a macho unilateral decision [by the vice-chancellor of the university].

The vice-chancellor confirmed this: 'This was not a decision based on consensus, but a decision made by the board. We invest in this unit. . . . Some things just have to be decided by the board.'

The third design strategy of *associative reasoning* was employed at two different stages. First, exemplars within the university and in direct proximity to the university are important sources of inspiration. For example, Programa Innova at UPC started to support the commercialization of intellectual property similar to the support already existing in the separate technology transfer unit of UPC. The same pattern is observed where policies and support methods were copied between institutions. For example, the independent Business Centre, located close to the WU campus, was the first unit to provide intensive support to new ventures. Later, WU decided to establish its own policies and support staff, which previously only consisted of a limited number of people involved in technology transfer. WU also decided to invest in companies, something previously done by one of the partner organizations of WU and the Business Centre. Conversely, the regional consortium, in which investors were also involved, wished to become involved in scouting ideas at WU departments, something that is currently only done by WBG. Similar local copying movements were observed at UPC. For example, another university in the proximity of UPC had developed a successful science park. The regional government agency identified this as a good example, motivating UPC to try to establish a similar science park. Similarly, the successful transformation of university culture at UPC apparently set an example for other universities in Spain.

Second, designers of new venture support also referred to more distant cases as sources of inspiration for designing and starting new venturing programs in universities. Widely reported new venture stories from elite US universities such as MIT and Stanford (Roberts 1991; Roberts and Malone 1996) led European university boards to pay more attention to university new venture stimulation, in addition to other means of technology transfer. For example, the director of Innovation Lab (TU/e) visited universities in Singapore and Stockholm, among others, and learned that adopting a good principle for the distribution of revenues was very important.

Another example is UPC, which adopted the idea of a business angel network from similar networks elsewhere in Europe. These analogies not only serve to design new policies and support activities, but also to justify them. Moreover, analogies with other cases served to identify dissimilarities. The vice-chancellor of WU stated:

> With the director of WBG, I made a trip to six top universities in the States. . . . We saw that it is important that the research group around the researcher profits from the commercialization. . . . We did not imitate this exactly, but we followed the same principles, in terms of distributing some of the financial benefits of a successful new venture to the research group.

In addition, such site visits facilitated the development of a common mindset among decision-makers on new venture strategies. However, personal biases and experience affected the associative reasoning raised by these visits. This is evident in the different way the vice-chancellor of WU and the director of WBG reflected on their visit to American universities. The director of WBG saw that 'as a university, it is important to have money at your disposal to invest. Most universities completely depend on funding from third parties. That undermines your position in negotiations with these external parties'. Therefore, the director of WBG developed a fund of several million euros for large investments in new ventures on behalf of the university. In contrast, the vice-chancellor of WU concluded that 'the philosophy of generating rigid income by large investments in companies by us . . . is a very risky business. We should be very careful with this! If this is a separate fund and it is directing the research then we are pushing the wrong buttons'.

So, the use of analogies to a large extent depends upon the designer's background and experience. Furthermore, in the case of large differences between semi-autonomous departments (e.g. WU), analogies not only result in new, shared perspectives but also in diverging interpretations. Hierarchy is especially related to the ability to implement the observed analogies and to communicate this as an inspiring vision, although this is moderated by the local culture of cooperation. The director of Innovation Lab at TU/e stated: '[In Singapore:] if the boss says: "to the left", we go to the left. But here, we keep talking for 4–5–6–7 years! . . . However, here it is easier . . . Just walk into the office.'

The use of more distant exemplary cases, such as MIT and Stanford, appears to be related to the international character of the university, and as such is more frequently observed at WU and TU/e than at UPC. Local exemplars are noted slightly more in the case of UPC, probably due to close proximity to the regional government, which enforces regional convergence and interwovenness.

2.4.2 Design Strategies and Performance of the Design Process

The data discussed thus far suggest that the design of new venture support policies and activities draws mainly on associative reasoning, whereas feedback-driven learning is also important at UPC and TU/e and off-line reasoning and planning plays a significant role at WU and TU/e. The importance of off-line reasoning and planning can be illustrated by quoting the director of WBG, who described the sources for his redesign planning process as follows:

> The board said: wouldn't you like to think about setting up an organization that forms the 'corporate' valorization? . . . I got a room in this building and I started to analyze. . . . I started to think about how this world [of new venture support and technology transfer] looks: what are success and failure factors, what is the chance of success? Once I mapped this, I thought about what are the factors a knowledge institute can contribute to make new ventures successful; to maximize the chance of success. I did this by talking to a lot of people, by reading a lot, and by looking around.

Moreover, our findings indicate that the use of the design strategies and their contribution to the performance of the design process is strongly associated with specific context characteristics. The context influences the degree to which one of the design strategies will result in the desired outcome – a successful and viable new venture unit. The context also affects the easiness of communicating and implementing the design efforts; in other words, the context also influences the efficiency of each of the design strategies. Our data suggest that hierarchy and near-decomposability in terms of money streams are important determinants of the contribution of design strategies to the performance of the design process. In this respect, the WU case illustrates how some of these context characteristics surfaced in a very explicit way, relating to tensions about power and ownership of revenues and success. Referring to the role of the board and management team at WU, the director of WBG argued:

> I think that an important part of the discussions involved the question of [political] power. They were of the opinion that WBG had become too powerful. That's the reason they killed it. . . . I was the only one who had an independent budget within this organization. Everybody was jealous.

In retrospect, these tensions resulted in a major redesign, which in fact implied the end of the WBG unit in its old form. The lack of feedback-driven learning resulted in limited anchoring of the unit in the nearly decomposable semi-autonomous departments, which in turn led to low effectiveness of the design process, in this case in terms of survival.

Table 2.3 summarizes our main findings by specifying the normative implications of the design strategies in terms of design results (in particular arriving at the intended results and the viability of the design results) and efficiency of the design process in terms of effort and time needed. This table is structured in terms of context characteristics: near-decomposability, hierarchy and interaction with regional and national policymakers. Concerning the university as a complex organization, we found that the degree of loose coupling of departments and hierarchy shape the use of different design strategies and affect the associated performance of the design process. For instance, in an organization with high near-decomposability, feedback-driven learning is needed to anchor the design in academic departments, thus increasing the survival chances of the designed unit. In terms of efficiency, however, feedback-driven learning is time-consuming and can, in different parts of the organization, result in divergent feedback. For the policy context, the efficiency and results of design strategies are largely influenced by the intensity of interaction with regional and national policymakers. Here, especially off-line reasoning and planning appear to be instrumental in justifying and anchoring the design as well as communicating the design process.

2.5 DISCUSSION AND CONCLUSION

2.5.1 Theoretical Implications

Our empirical study explores how designers of organizations use different (cognitive) design strategies in different contexts. The data from three cases suggest that contextual conditions influence the use of these strategies and affect the associated performance of the design process (see Table 2.3). Our findings indicate that associative reasoning prevails in the case contexts and that feedback-driven learning is instrumental in anchoring the designs to enhance the viability of the designed units. In this respect, this study underscores earlier theoretical claims that designers in moderately complex and novel settings preferably engage in associative reasoning by way of analogies (Gavetti et al. 2005, 2008). Moreover, this study replicates the finding that both experimentation and analogies (or copies) are very important to the design of separate yet bridging new venture units in a complex organization (Hill and Birkinshaw 2008).

Most importantly, our findings serve to specify *how* organization designers use these strategies and combinations of these strategies to design new venture units. Our findings also serve to specify the performance of the design process associated with employing different strategies in diverse

Table 2.3 Design strategies in relation to context characteristics

Context characteristics		Off-line reasoning and planning	Feedback-driven learning	Associative reasoning
Organization				
Near-decomposability	High	R[a]: *medium*, not anchored E[b]: *high*, easy justification	R: *high*, anchoring the design E: *low*, potentially diverging and time consuming	R: *high*, avoiding lock-in E: *medium*, guiding but potentially diverging
	Low	R: *medium*, not anchored E: *high*, easy communication	R: *high*, anchoring E: *medium*, time consuming	R: *high*, avoiding lock-in E: *high*, converging
Hierarchy	High	R: *medium*, threat of resistance E: *high*, convince and justify	R: *medium*, difficult to anchor at all hierarchical levels E: *medium*, difficult to learn from all hierarchical levels	R: *high*, justified design E: *high*, easy communication
	Low	R: *medium*, not anchored E: *low*, difficult to communicate	R: *high*, fine-tuning design E: *medium*, dispersed interaction	R: *high*, justified design E: *low*, difficult to communicate
National and regional policy interactions		R: *high*, justified design E: *high*, easy to communicate	R: *high*, foster convergence E: *low*, time consuming	R: *high*, justified design E: *medium*, dependent on relationships

a R: the importance of this design strategy for design results/effectiveness: arriving at the intended results (i.e. a new venture unit supporting new venture creation) and the viability of the design results (i.e. survival of the unit) in this particular context.

b E: the importance of this design strategy to efficiently perform the design process, in terms of the efforts and time needed, in this particular context.

contexts (characterized by near-decomposability, hierarchy and interaction with policymakers, Table 2.3). These insights in the context-specific contribution of the design strategies substantially extend previous research on the design of new venture units (Hill and Birkinshaw 2008) and on organization design (Gavetti et al. 2008). In this respect, this study underlines that the design strategy of associative reasoning creates the ability to acknowledge differences between the situation at hand and the exemplar situation, which tends to result in high performing design processes (see Table 2.3). As such, an analogy can function as a mental framework that incorporates (and possibly integrates) different aspects of the object being designed. In the design efforts and implementation of support activities and policies, a particular exemplar – for example, the venturing system at MIT – can provide a powerful vision to avoid lock-in and to justify the design.

This study also provides a number of new insights by linking the associative reasoning strategy to the design context. First, the degree to which analogies can be used efficiently apparently depends on the hierarchy of the organization and the place of the designers in this hierarchy. Moreover, the viability of the design result in the case of associative reasoning is subject to the complementary use of the strategy of feedback-driven learning, as the WU case shows. Furthermore, the empirical findings suggest that associative reasoning is used at several levels in and around the organization, which can then lead to converging mindsets around a particular exemplar. However, in the case of conflicting interests of the decision-makers, the exemplar situation can also create divergence (e.g. at WU).

Another finding that extends previous theorizing is that feedback-driven learning may also prevail in a moderately complex and novel context, and not merely in situations characterized by extreme novelty and high complexity. Feedback-driven learning is especially instrumental in adapting given (preliminary) designs to the context and anchoring them in the organization. As all three universities are to a large extent nearly decomposable, feedback-driven learning appears to enable the designers to create a new venture support system that is tailored to the needs of particular lower hierarchical levels (departments, research groups, individual faculty). However, the UPC case shows that a high degree of hierarchy tends to limit feedback-driven learning to only higher levels of the hierarchy, thus highlighting a risk that is not acknowledged in previous research. Moreover, in the case of WU's central new venture support organization that depends highly on loosely coupled departments, feedback-driven learning at the level of the support unit diverged from what was learned at the level of departments and thus tensions arose.

Our findings regarding the use of off-line reasoning and planning extend previous work in this area (Gavetti et al. 2005, 2008). We observed that all design strategies are employed, but on diverse hierarchical levels, in different degrees and with diverse results. Farjoun (2008) argued that off-line reasoning and planning is most likely to be adopted in familiar, stable and less complex settings. However, our study suggests that designers also use off-line reasoning and planning in moderately complex and moderately novel settings. In particular, this design strategy helps to 'sell' particular ideas to stakeholders and providers of resources, whereas the actual design process to a large extent draws on associative reasoning and feedback-driven learning. This finding may also serve to explain why some design studies emphasize feedback and association (e.g. Gavetti et al. 2005; Schön 1984), whereas other studies focus on planning and forecasting (e.g. Simon 1996). As such, our study demonstrates that all three design strategies are likely to be used in one empirical setting, but with different degrees of importance, at different organizational levels, and with differential impact on the performance of the design process.

The experience of agent-designers surfaces as an important designer-specific factor in dealing with contexts in all three cases. For example, at TU/e and WU, the directors of the new venture support organizations had previously acquired experience in designing similar organizations in other settings. Experience influences what can be done, which is apparently different at WU and TU/e: at WU, the designer chose to create a separate fund, while at TU/e, the director emphasized embeddedness in departments. As such, experience does not provide an exclusive explanation, but does help to understand how designers interpret analogies and integrate design knowledge from different sources. This is in line with previous work by Gavetti et al. (2005), who observed that the experience of the agent applying the analogy is an important predictor of the performance of the application in practice. In this respect, experienced designers tend to pick better analogies and are also likely to be more capable of adapting their analogies and strategies. At UPC, the use of analogies was observed to a lesser extent and feedback-driven learning was more important than in the two other cases. This may be a result of the incumbent designer having less business and international experience (Chi et al. 1981); moreover, this designer was evidently more integrated in the university's political system, as a former vice-president and dean. In this respect, experience of the designer(s) also seems to be an important moderator between the context and design process performance associated with using feedback-driven learning. The importance of the experience of designers highlights that design theory needs to carefully balance the role of designers and the role of specific design strategies codified in design principles.

2.5.2 Practical Implications

Our study has important implications for practitioners who engage in the (re)design of universities or other complex social systems, in particular when they are involved in incorporating somewhat foreign elements in the organization such as new venture units. First, our case studies illustrate the pivotal role of agent-designers in crafting separate new venture units, while interacting with the complexities of the organization. Designers are agents who (can) create coherence in a design and decide which design strategies to follow. Their experience provides a repertoire of potential analogies to be used in associative reasoning and helps in the evaluation of these analogies as well as the adaptation of design principles to a local context. Second, the study implies that in particular the design strategy of associative reasoning can be used to integrate the different design principles into one coherent design solution while being able to acknowledge differences between the situation at hand and the reference case.

The practical execution of this principle can include site visits to major exemplars or managers from one of these benchmark cases visiting the university. These visits can be the source of insights required for the associative reasoning strategy. Besides other universities, analogies could also include very different situations, such as successful enterprises. Third, the design strategy of feedback-driven learning should be used to adapt existing design blueprints – or if available, design principles and solutions – to the context and anchor them in the different layers of the university organization. As feedback-driven learning is very context-dependent, this includes inviting and using feedback from different people in the organization and its context, thus involving decision-makers, staff members and opinion leaders.

In sum, this study contributes to the innovation management and academic and corporate entrepreneurship literature by exploring the interaction between (re)design processes of separate new venture units and complex design contexts. Moreover, this study contributes to the organization design literature by describing the use of different design strategies in these specific contexts as well as specifying the contributions of these strategies to the performance of the design process. The design of new venture units in universities illustrates how difficult it is to deliberately change an organization with multiple linkages, a long history and semi-autonomous departments that are loosely coupled. Our study also points to the role of organizational near-decomposability, which forms a contextual characteristic that in interaction with hierarchy severely influences the domain available to designers and the applicability of top-down directives. In this respect, the case studies in this chapter emphasize the power of associative reasoning as well as the need for feedback-driven learning at different organizational levels, and by people with certain levels of experience. Overall, our study suggests that design strategies need to be understood in relation to the complexity of the context in which they are adopted.

REFERENCES

Ambos, T.C., K. Mäkelä, J. Birkinshaw and P. D'Este (2008), 'When does university research get commercialized? Creating ambidexterity in research institutions', *Journal of Management Studies*, **45** (8), 1424–47.

Bekkers, R., V. Gilsing and M. van der Steen (2006), 'Determining factors of the effectiveness of IP-based spin-offs: Comparing the Netherlands and the US', *Journal of Technology Transfer*, **31** (5), 545–66.

Chi, M.T., P.J. Feltovich and R. Glaser (1981), 'Categorization and representation of physics problems by experts and novices', *Cognitive Science*, **5** (2), 121–52.

Ethiraj, S.K. and D. Levinthal (2004), 'Bounded rationality and the search for organizational architecture: An evolutionary perspective on the design of organizations and their evolvability', *Administrative Science Quarterly*, **49** (3), 404–37.

Farjoun, M. (2008), 'Strategy making, novelty and analogical reasoning: Commentary on Gavetti, Levinthal, and Rivkin (2005)', *Strategic Management Journal*, **29** (9), 1001–16.

Gavetti, G., D.A. Levinthal and J.W. Rivkin (2005), 'Strategy making in novel and complex worlds: The power of analogy', *Strategic Management Journal*, **26** (8), 691–712.

Gavetti, G., D. Levinthal and J.W. Rivkin (2008), 'Response to Farjoun's Strategy making, novelty, and analogical reasoning – commentary on Gavetti, Levinthal, and Rivkin (2005)', *Strategic Management Journal*, **29** (9), 1017–21.

Hill, S.A. and J. Birkinshaw (2008), 'Strategy-organization configurations in corporate venture units: Impact on performance and survival', *Journal of Business Venturing*, **23** (4), 423–44.

Jansen, J.J.P., M.P. Tempelaar, F.A.J. van den Bosch and H.W. Volberda (2009), 'Structural differentiation and ambidexterity: The mediating role of integration mechanisms', *Organization Science*, **20** (4), 797–811.

Lawson, B. (2006), *How designers think: The design process demystified* (Amsterdam: Architectural Press).

Roberts, E.B. (1991), 'The technological base of the new enterprise', *Research Policy*, **20** (4), 283–98.

Roberts, E.B. and D.E. Malone (1996), 'Policies and structures for spinning off new companies from research and development organizations', *R&D Management*, **26** (1), 17–48.

Romme, A.G.L. and G. Endenburg (2006), 'Construction principles and design rules in the case of circular design', *Organization Science*, **17** (2), 287–97.

Schön, D.A. (1984), *The reflective practitioner: How professionals think in action* (Aldershot: Ashgate).

Schön, D.A. (1987), *Educating the reflective practitioner: Toward a new design for teaching and learning in the professions* (San Francisco: Jossey-Bass).

Simon, H.A. (1962), 'The architecture of complexity', *Proceedings of the American Philosophical Society*, **106** (6), 467–82.

Simon, H.A. (1996), *The sciences of the artificial* (Cambridge: MIT Press).

Van Burg, E. (2010), *Creating spin-off: Designing entrepreneurship conducive universities* (Eindhoven: Eindhoven University Press).

Van Burg, E., A.G.L. Romme, V.A. Gilsing and I.M.M.J. Reymen (2008), 'Creating university spin-offs: A science-based design perspective', *Journal of Product Innovation Management*, **25** (2), 114–28.

Weick, K.E. (1990), *Introduction: Cartographic myths in organizations in mapping strategic thought*, A.S. Huff (ed.) (Chichester: Wiley).

3. TU Berlin – an entrepreneurial university in an entrepreneurial city

Matthias Mrozewski, Agnes von Matuschka, Jan Kratzer and Gunter Festel

3.1 BERLIN AS ENTREPRENEURIAL CITY

The regional economy of Berlin, the capital of Germany, is notably affected by structural transformation, mainly resulting from the shrinking of traditional heavy industry and the construction sector. Many of these traditional and labor-intensive industries no longer exist or have relocated their activities to low-cost countries. As a consequence, Berlin currently has an unemployment rate considerably higher than the average rate in Germany. At the same time, new industries are developing in Berlin, such as renewable energy, biotechnology, medical technology, microsystems technology, information and communication as well as traffic and mobility technologies. In the last ten years, they have been collectively responsible for the creation of tens of thousands of new jobs in the Berlin region. Future-oriented and innovation-driven, these industries are expected to be more sustainable than their traditional counterparts and to generate long-term economic, social and environmental value. This is because they create jobs for highly qualified employees and, unlike traditional industries, they rely to a lesser extent on natural resources and to a larger extent on human knowledge and efficient technology.

Further development of these industries depends on the availability of human capital and requires the strong presence of research institutions, particularly with a technical focus. In this respect, Berlin seems to provide very favorable conditions because it is home to several universities and applied sciences universities with nearly 140,000 students. Major research institutions, such as the Max-Planck-Society and Fraunhofer Society, are located in and around Berlin. The city also has two major technology parks, Campus Berlin Buch and Adlershof, the latter being one of the largest science and technology parks in the world. These institutions, however, represent only a fraction of Berlin's sophisticated scientific and economic infrastructure, which currently consists of nearly 600 public science and research related organizations (Berlin Sciences Navigator 2014).

Entrepreneurship and spin-off activities of students, research assistants, scientists or professors associated with these organizations are important. By founding companies, researchers introduce scientifically grounded technologies, innovative products and services or superior manufacturing processes into the economy (Festel 2013; Festel et al. 2014). In addition to technological advancement, there is also societal gain because those start-ups often grow into leading companies that create jobs and pay taxes. These entrepreneurial activities are powered by a distinct and vibrant founding culture in Berlin, the city with the most start-ups in Germany with regard to the economically active population. In 2009, more than 7,500 companies were founded in Berlin, which corresponds to 1,310 start-ups per 10,000 existing companies. Berlin also has concrete entrepreneurship support systems in place. The city financially supports promising start-ups as well as industry-specific centers for technology and entrepreneurship. The Berlin-Brandenburg Business Plan Competition is the largest of its kind in Germany. This competition not only rewards business ideas but also academic initiatives to nurture and enhance entrepreneurial spirit and the founding of new businesses.

3.2 TU BERLIN AS ENTREPRENEURIAL UNIVERSITY

The Technical University Berlin (TU Berlin) is one of the oldest and largest technical universities in Germany. Today, TU Berlin has seven faculties, five of which are related to engineering or natural sciences. There are approximately 30,000 students, 2,400 research assistants and more than 320 professors. It should therefore come as no surprise that TU Berlin is currently considered to be the university with the highest technological innovation potential in the entire Berlin and Brandenburg region. Technology transfer and entrepreneurship activities to commercialize new technologies are regarded as crucial elements of TU Berlin's overall strategy and the university's management encourages their further development. Many efforts have been made in recent years to support the entrepreneurial activities of TU Berlin's students and academic staff.

In the 1970s, the university opened Germany's first center to support founders, which was later extended to the Berlin Technology and Innovation Park. To date, nearly 350 enterprises, most related to TU Berlin, have chosen to set up business there. In 2007, the Founder's Service, headed by Agnes von Matuschka, was established as one element of technology transfer within TU Berlin. The Founder's Service is responsible for providing practical support to TU Berlin entrepreneurs. It is a one-stop-shop for students, entrepreneurs and investors and provides them with entrepreneurship awareness workshops, consulting services, soft skills seminars as well as office space in its incubator. The Founder's Service supports potential founders at every stage of the start-up project, regardless of whether these are first ideas or a fully developed business plans. The goal of the Founder's Service is to enhance entrepreneurial thinking among students and researchers and advise them during the founding and technology transfer process. New sensitization and qualification methods have been developed to make students, graduates and research assistants more aware of self-employment as a professional alternative. The Founder's Service provides online tools, e.g. a test to measure entrepreneurial aptitude, and potential founders are able to properly evaluate the pros and cons of being an entrepreneur during a special workshop hosted by a psychologist.

The Center for Entrepreneurship (CfE) was established in 2010 and is operated by the Founder's Service at TU Berlin together with the Chair of Entrepreneurship and Innovation Management headed by Professor Jan Kratzer. Established in 2009 and funded by Siemens, this Chair is part of the Institute for Technology and Management within the School of Economics and Management at TU Berlin and is responsible for the research perspective of the CfE. The Chair strives to generate internationally respected research and share relevant knowledge with students, scholars, potential entrepreneurs, politicians and the corporate world. The Chair is clearly research-driven and its primary objective is to contribute to literature through contemporary, conceptually innovative, managerially relevant and internationally recognized publications. The Chair's research agenda is driven by empirical and practical relevance and its ultimate objective is to provide guidance to entrepreneurs, innovators and managers on how they can improve their management practices.

For its research projects, the Chair works closely with the Chair of Technology and Innovation Management and frequently cooperates with the Institute of Technology and Management at TU Berlin as well as with researchers from institutions around the world. The Chair's international orientation and interdisciplinary approach is reflected in the composition of its team, which currently includes researchers from China, France, Germany, Poland, Switzerland, Tanzania and Ukraine. Their academic backgrounds include engineering, finance, geography, management, sociology and psychology. All team members have several years of industry, consulting or start-up experience. In addition to conducting state-of-the-art research, the Chair provides students with a broad range

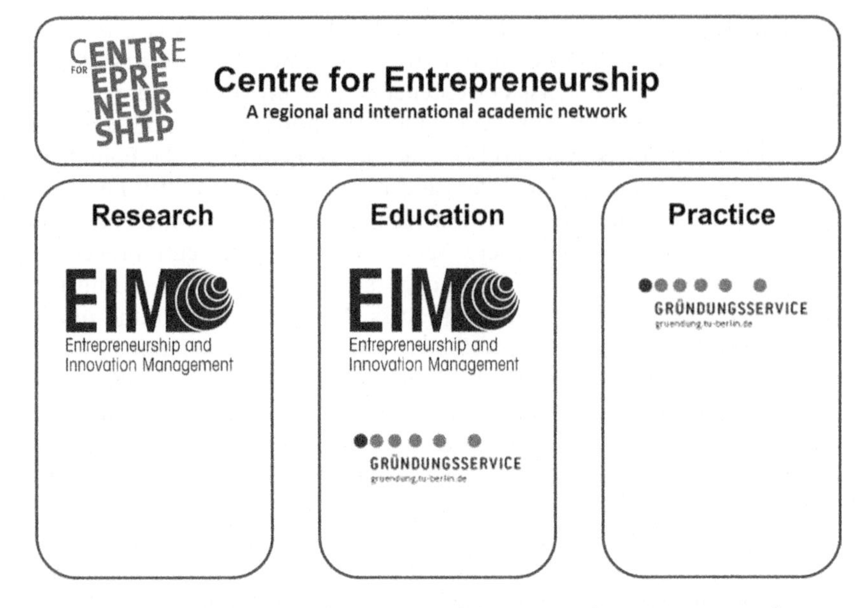

Figure 3.1 Organizational structure of the Center for Entrepreneurship

of entrepreneurship training, from scientifically grounded entrepreneurship research courses to practice-oriented soft skills and business plan seminars.

The activities of the CfE are divided into three main areas: research, education and practice (Figure 3.1). This organizational structure makes it possible to bundle all entrepreneurship-related competences at TU Berlin. Although all universities in Berlin systematically promote and support spin-offs and undertake important tasks in the field of technology transfer, TU Berlin stands out from the crowd. This is confirmed by its recent success in a nationwide competition for the most entrepreneurial university in Germany. TU Berlin, with its CfE, came top out of 83 universities and applied sciences universities, and is now entitled to use the prestigious label 'Entrepreneurial University'. TU Berlin's performance is also demonstrated by its attaining second position in the latest Schmude Ranking from 2011, which compares 59 German universities and ranks them in different categories according to the effectiveness of their entrepreneurship support systems (Schmude et al. 2011). Since 2007, TU Berlin start-ups successfully pass the application process for EXIST scholarships for founders and research transfer, granted by the German Ministry of Economy and Technology to excellent start-up teams from universities in order to support them in the foundation and start-up stage. This support amounts to more than 3.5 million euros in total and TU Berlin belongs in the top five German universities with the highest number of EXIST foundation scholarships. TU Berlin's effectiveness in entrepreneurship support has even been recognized internationally. In 2009, the Organisation for Economic Co-operation and Development (OECD) selected TU Berlin as a best practice example for entrepreneurship support systems in its international study on entrepreneurship and innovation at universities.

3.3 EDUCATION PROGRAMS IN ENTREPRENEURSHIP

The CfE provides state-of-the-art education to its students, guest students from other Berlin universities and guest auditors practicing in the field of business administration and particularly entrepreneurship and innovation management. Thereby, it tries to bridge the gap between academics and practitioners by providing academically ambitious lectures on research problems in entrepreneurship as well as practical, business-oriented lectures on business planning, company founding and start-up management. For example, the theoretically oriented entrepreneurship research course aims to give students a 360-degree overview of all research streams around the theme 'entrepreneurship', from managing and aligning internal and external networks, to every facet of founding an enterprise and how to explore and develop markets. The course studies both theoretical models and their empirical application, alternating between microeconomic and macroeconomic perspectives with a strong methodological focus.

In co-operation with Twente University in the Netherlands, the double Innovation Management and Entrepreneurship Master's program has been established. This course is a unique and exclusive double degree combining the interrelated disciplines of innovation management and entrepreneurship in one program. Its international focus attracts a high number of foreign students. The program equips students with in-depth knowledge of innovation management on a strategic, tactical and operational level, as well as all aspects of entrepreneurship. Lectures are delivered by academics from different university institutes as well as experts from politics and industry. Participants also get access to a network of companies associated with the program including Akzo Nobel, Deutsche Telekom, Philips, Siemens and ThyssenKrupp.

Another course called Venture Campus offers students a series of practical lectures on business planning and start-up founding. The aim is to support and realize the high-tech start-up potential of TU Berlin students.

Venture Campus provides students with all the business and management skills necessary for launching a start-up company and covers all aspects of the founding process. During the course, students work closely together in interdisciplinary teams to prepare business plans and simulate the founding of a company. Since its launch in 2004, approximately 500 students have completed this course and 150 business plans were prepared. Approximately 10% of these translated into real start-ups. The most popular sectors were retail, internet-based services and services in advertising and real estate.

Another important focus of the CfE is skills development. Future founders are prepared for their role by developing professional and commercial know-how as well as personal and social skills training. The Founder's Service offers workshops and seminars to take these skills to the next level by addressing the following topics: first steps into self-employment, negotiation skills, corporate design, project management, procurement and presentation techniques. Another training opportunity is the one-week Entrepreneurship Academy, attended by 20 selected founders from technology-oriented domains. The intensive program comprises workshops given by experienced legal, tax and management consultants, business coaches, entrepreneurs and potential investors. This unique approach combined with the high quality of lecturers makes the Entrepreneurship Academy application process highly selective. Potential participants must be entrepreneurs from the Berlin region who have founded a start-up in the last 12 months. Applicants must send their CV, a short presentation of their start-up and a motivation statement explaining why they want to participate.

The development and use of dormant innovation potential is achieved by the early evaluation of most TU research projects from a market-oriented perspective. Any gaps identified in the innovation process are then addressed by corresponding support instruments. The CfE offers a free technology-screening workshop called Product Propeller to research assistants. The potential of research results is not always visible at first sight. By providing them with different perspectives on their research results, the workshop helps research assistants to identify and assess market potential, and serves as a platform for the development of innovative ideas for product and services based on their research. Scientific insights are studied from a market-oriented perspective using creative techniques. This workshop is contributing to the establishment of realizable ideas, and a shift towards a more market-oriented way of thinking, supported by further guidance on start-up founding as well as detailed information on potential sources of finance. The success of this workshop stems from the fact that all team members are treated equally and work and generate ideas together, which are then clustered and assessed. Although the process is very structured by a moderator, it still leaves space for innovative ideas to emerge.

Another very recent project is Program Plan B, which is specifically designed to meet the needs of potential female entrepreneurs. The target group is female graduates who are thinking about or planning to become entrepreneurs but also professional women who want to change career paths. In both cases, applicants must have a university degree and they need to apply formally. The four-month program includes advice from female academics on why participants should consider entrepreneurship as an interesting career alternative. Twenty participants are chosen during a highly selective application process and then assigned to small teams where they work together to develop new business ideas, create a business plan and start preparations for founding a company. At the end of the course, the best business plan receives an award and the winning team has the possibility to use the incubation facilities of the Founder's Service for one year, free of charge.

3.4 PRACTICAL SUPPORT FOR FOUNDERS

TU Berlin organizes regular presentations of successful start-ups in order to share best practices. Every six months, a special open day offers founders the opportunity to participate in workshops and discussions, and to get to know successful entrepreneurs. The traveling exhibition 'Show your profile' also features several successful founders from TU Berlin. In 2014, the number of entrepreneur profiles in this exhibition reached 100. Together with a growing catalogue of former TU Berlin founders, this exhibition communicates the considerable entrepreneurial potential of the university and is helping to convince potential entrepreneurs to choose this alternative career path. Another initiative aimed at increasing students' awareness of entrepreneurship is an information day organized by the Founder's Service, which consists of lectures and workshops on the entrepreneurship support systems available at TU Berlin plus business modeling.

Another important activity is providing advice on how to improve business planning to support business ideas, an essential ingredient of a successful presence on the market. Experienced start-up consultants provide help or advice on start-up projects. This includes assistance with the elaboration of a business plan, counseling on funding and financing (e.g. federal financial support, such as the EXIST scholarship for founders or the EXIST research transfer) as well as establishing contacts with business angels and venture capital investors. If a team of potential founders lacks partners it can use the online partnership search engine to find them.

With its Founding Factory, the Founder's Service provides a professional business incubation service with 13 offices plus meeting rooms as well as a library with relevant literature for entrepreneurs. Consultants and other start-ups are located in the same building and this creates a fertile environment for sharing ideas. Due to limited space, teams are carefully selected. In the first stage of the selection process, interested teams present a concrete business idea in the form of an idea sketch. Having successfully passed the first stage, they either have to pitch their business idea to the Founding Factory selection committee or provide written confirmation that they have received an EXIST scholarship. Once a team has successfully passed both stages, they get access to the business incubation service for free for 12 months. After establishing their business, teams can still use equipment, laboratories and offices at the university.

3.5 REGIONAL AND INTERNATIONAL NETWORKING

Regional networking activities reflect the increased cooperation between TU Berlin and similar institutions in the Berlin region. Via the B!Gründet network, the CfE is connected with 12 other member organizations, including all universities and universities of applied sciences in Berlin as well as Berlin-based innovation centers and technology marketers. Within this network, joint marketing events are developed, for example participating in national Entrepreneur Days.

TU Berlin also actively networks with its former students. Most of the 900 TU alumni, who have already founded an enterprise, are willing to share their knowledge and experience as consultants, mentors or best-practice examples with others considering starting up their own business. Regular alumni angel evenings bring together young start-ups from TU Berlin and former graduates of TU Berlin. During these events, selected TU Berlin start-ups present their business ideas to invited alumni, who then award a Start-Up Label to the best concept. The first alumni angel evening took place in October 2011. Ever since, these events have provided budding entrepreneurs with an opportunity to exchange experiences, network with potential partners, acquire new customers and establish useful business contacts or initiate strategic alliances in a very short time.

TU Berlin is also actively engaged in international cooperation projects. These activities result from the belief that the increasing internationalization of entrepreneurship will bring about huge changes in entrepreneurship research, teaching and support facilities across Europe. TU Berlin takes a leading role within the Knowledge and Innovation Communities

(KICs) of the newly established European Institute of Technology (EIT), a pan-European alliance of leading European technical universities. The KICs bring together key actors in the innovation space such as companies, entrepreneurs, research and technology organizations, higher education institutions, investment communities, research funders and local, regional and national governments. Together with other major European innovation hubs, TU Berlin is a co-location center for two KICs, namely the ICT Labs KIC and the Climate KIC.

The ICT Labs KIC aims at the radical transformation of Europe into a knowledge society with an unprecedented proliferation of internet-based services. The consortium covers five major locations (Berlin, Eindhoven, Helsinki, Paris and Stockholm) and connects world-leading companies, globally renowned research institutes and top-ranking universities in those cities, all dedicated to speeding up innovation in order to address society's greatest challenges. The consortium covers the entire value chain in the ICT sector, thereby facilitating the capture of new business opportunities. Committed to an efficient open innovation model, the KIC enables faster transformation of ideas and ICT technologies into real products and services. It catalyzes the creation of strong ventures and helps them to grow to become the future world leaders in the ICT arena. TU Berlin plays a major role in the ICT Labs KIC through several educational initiatives and projects focusing on improving entrepreneurship support systems in Europe. Together with its KIC partners, TU Berlin is working on a new financing model for young high-tech start-ups in the seed stage. TU Berlin uses its pan-European network to improve the financing possibilities for regional start-ups by organizing international pitching events for high-tech start-ups from the KIC partner nodes.

The Climate KIC was launched to drive innovation in the area of climate change adaptation and mitigation through an integrated European network of global and regional partners from business, government and universities. The Climate KIC focuses on achieving excellence in four domains: assessing climate change and managing its drivers, transitioning to low-carbon resilient cities, adaptive water management and zero carbon production. These domains were selected for their mitigation/adaptation potential as well as innovation and job creation potential. Each domain has a world-renowned lead responsible for the entire chain from education to commercialization. Work is embedded in the five locations, with each location responsible for creating a local innovation ecosystem to drive entrepreneurship and venture creation, while linking into a network of implementation sites. Together with London, Zurich, the Paris metropolitan area and the Randstad metropolitan area, TU Berlin and its corporate partners form the Climate KIC co-location center.

REFERENCES

Berlin Sciences Navigator (2014), 'Science Navigator', http://www.berlin.de/sciencesnavigator (accessed 15 January 2016).

Festel, G. (2013), 'Academic spin-offs, corporate spin-outs and company internal start-ups as technology transfer approach', *The Journal of Technology Transfer*, **38** (4), 454–70.

Festel, G., Breitenmoser, P., Würmseher, M. and Kratzer, J. (2014), 'Early stage technology investments of pre-seed venture capitalists', *International Journal of Entrepreneurial Venturing*, **7** (4), 370–95.

Kirchner, U. and von Matuschka A. (2009), *TU-Gründungsumfrage 2008/2009* (Berlin: TU Berlin).

Schmude, J., Aevermann, T. and Heumann, S. (2011), '*Vom Studenten zum Unternehmer: Welche Universität bietet die besten Chancen?*' (Munich: LMU München).

PART 2

Supporting and coaching spin-offs

4. Systematic spin-off processes in university–industry ecosystems

Helmut Schönenberger

4.1 INTRODUCTION

This chapter discusses a new systematic spin-off process that has been developed and piloted since 2002 by UnternehmerTUM, the center for innovation and business creation at Technische Universität München (TUM). The objective of this university-affiliated entrepreneurship center is to initiate and accelerate more innovative companies that will develop into large global players. Munich, with its ongoing economic success, its cluster of mid-sized and large companies, its entrepreneurial history and its recently expanding culture of entrepreneurial endeavor, provides a highly supportive ecosystem for the development of scalable start-ups. Nevertheless, Munich has not been able to regain the momentum of the high-growth 'Wirtschaftswunder' period in the 1950s and 1960s. By constantly building and improving a new spin-off process, TUM and UnternehmerTUM have shown new ways to reignite a substantial flow of university-related start-ups. In 2013, more than 20 scalable high-tech start-ups were founded within the network.

This chapter starts with an overview of regional characteristics, the entrepreneurship strategy of TUM, and the setup of its entrepreneurship center, UnternehmerTUM. In the second part, the implemented spin-off process is explained in more detail, followed by a description of KICKSTART, the incubation program of UnternehmerTUM and TUM. Around 80% of all TUM start-ups focus on business-to-business markets. This means that the entrepreneurship center is also specialized in initiating partnerships between start-ups and established companies. The enabling activities to generate more start-up-industry collaborations are the focus of the third part. The chapter ends with a discussion of current challenges and the future outlook.

4.2 MUNICH ECOSYSTEM, TUM – THE ENTREPRENEURIAL UNIVERSITY AND UNTERNEHMERTUM

Munich has grown to become one of the leading high-tech clusters in Europe. The region is home to a large number of mature corporations, small and medium-sized enterprises, universities, research and development centers and various service providers. Munich experienced phases of exceptional economic growth, especially in the second half of the nineteenth century and after World War II. During these periods, companies like MAN, MunichRe, Allianz, Siemens and BMW developed into leading corporations and became major global players. Despite its entrepreneurial history and ongoing economic success, Munich's ecosystem has not produced a large number of innovative high-growth companies in recent decades compared with other innovation clusters like Silicon Valley (see also Engel, 2014).

To face this challenge, TUM, which is the alma mater of world-famous innovators and entrepreneurs like Carl von Linde, Rudolf Diesel, Claude Dornier and Willy Messerschmitt, has dedicated itself to becoming 'The Entrepreneurial University' and has developed a strategic innovation program called TUM Entrepreneurship. The central goal of this university-wide action plan is to initiate more scalable high-tech start-ups. TUM is one of the top-ranking research universities in Europe with 13 departments, 154 degree programs and around 36,000 students, 22% of whom come from outside Germany. The faculty portfolio comprises engineering, natural sciences, medicine and life sciences as well as management. TUM academic departments have formed many interdisciplinary research groups and institutes that are often the birthplace of disruptive innovation. The total annual budget of the university exceeded 1.2 billion euros in 2013.

TUM Entrepreneurship focuses especially on PhD students and faculty members in order to commercially exploit research results and create high-growth spin-offs. The initiative is funded by the German Ministry of Economics and Technology (BMWi) within the national program 'EXIST-Culture of Entrepreneurship' and is implemented by TUM Entrepreneurship professors, the technology transfer office, TUM ForTe, and its entrepreneurship center, UnternehmerTUM, which serves as an enabler for high-growth ventures. TUM Entrepreneurship is dedicated to initiating as well as accelerating spin-off companies by enhancing the systematic innovation and business creation process that bridges the gap between the university and industry. In order to address the needs of science and business communities, its entrepreneurship center is set up both as an affiliated university institute and as a privately owned company.

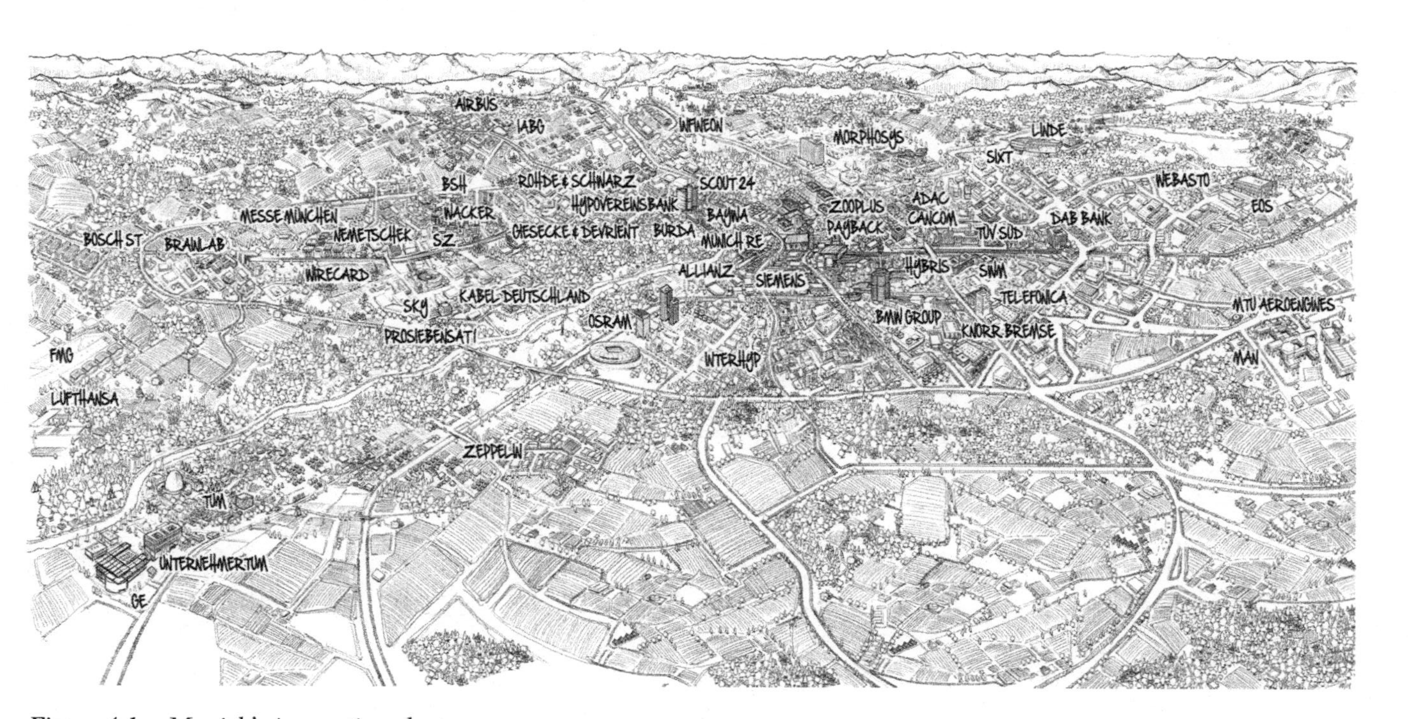

Figure 4.1 Munich's innovation cluster

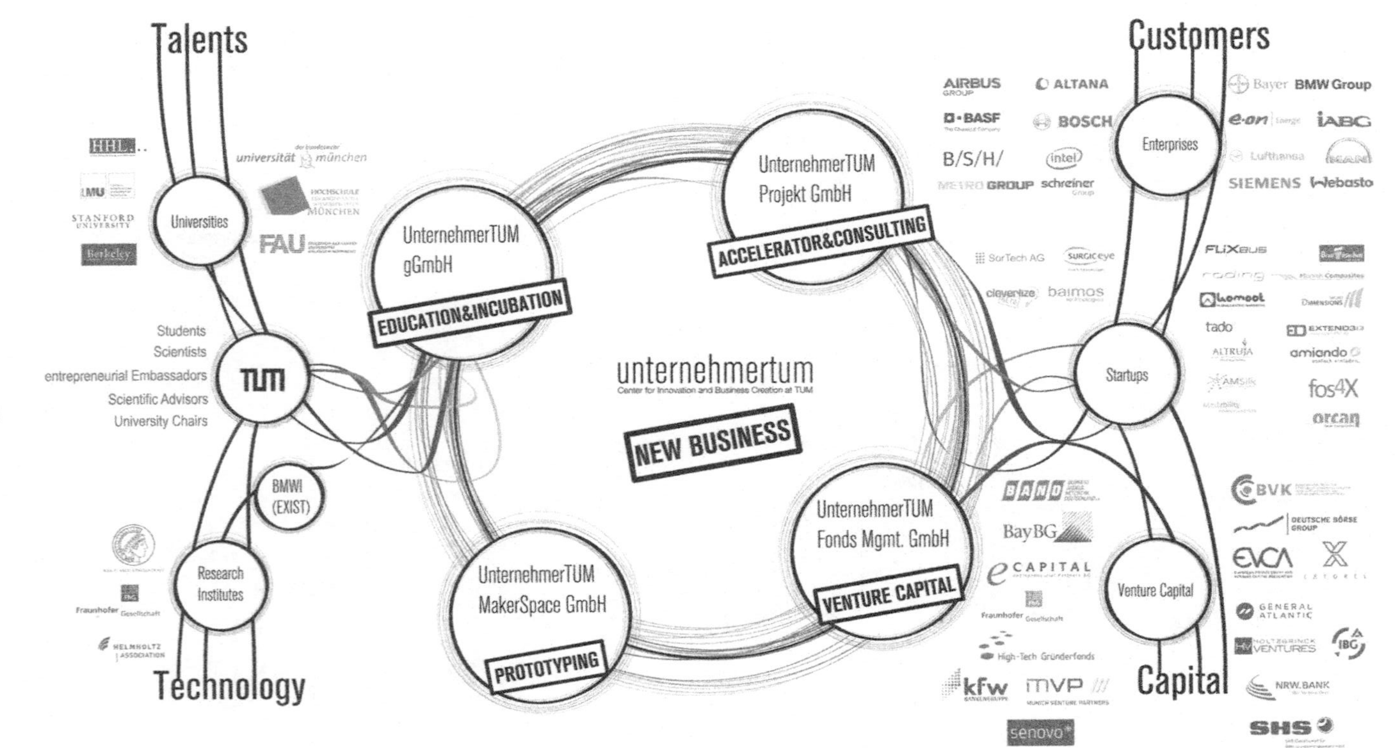

Figure 4.2 Structure and ecosystem of UnternehmerTUM

Founded in 2002, UnternehmerTUM has grown into one of the leading university-based entrepreneurship centers in Europe with more than 1,000 participants attending lectures, seminars and programs as well as more than 50 innovation and start-up projects every year. In 2013, more than 20 scalable high-tech start-ups evolved out of this network of activities. In order to serve as an entrepreneurial role model and to generate revenue for its own sustainability and evolution, the center is structured as a private company owned by Susanne Klatten, a key German entrepreneur and investor.

With more than 100 employees, UnternehmerTUM is divided into four companies. For tax and corporate governance reasons, UnternehmerTUM GmbH has bundled its for-profit business into three subsidiaries: UnternehmerTUM Projekt GmbH, UnternehmerTUM-Fonds Management GmbH and UnternehmerTUM MakerSpace GmbH.

The non-profit UnternehmerTUM GmbH, supported to a large extent by donations from Susanne Klatten, focuses on all education, networking and university early-stage spin-off activities. Its main task is to inspire and educate students, academics and professionals to become entrepreneurs and to support the creation of technology-based university start-ups. UnternehmerTUM GmbH also incubates and accelerates technology-based start-up teams, supplying them with coaching, mentoring and access to offices and a prototyping infrastructure.

The second company, UnternehmerTUM Projekt GmbH, manages the for-profit acceleration programs as well as the innovation consulting business. Together with clients from start-ups and established companies like Airbus, BASF, BMW, EON, MAN and Siemens, project teams identify opportunities, develop business models and build start-ups and new business units for existing corporations. One example is a spin-off project with Airbus, which is commercializing a rapid microbiological water analysis bio-detector invented by its corporate research department. Consulting projects also help to engage a broader business and entrepreneurial ecology in the TUM sphere. Some of the projects are staffed with students from entrepreneurship courses who benefit tremendously from this experimental learning experience.

UnternehmerTUM-Fonds Management GmbH, the third company, runs a venture capital fund to finance cutting-edge technology start-ups with international market potential. The 25 million euro fund is structured like most other privately held, independent venture capital companies. Around a quarter of the companies in its portfolio are directly related to TUM. Most of the other start-ups have their origins in other German universities and research organizations.

MakerSpace GmbH, the fourth company of UnternehmerTUM, is a

1,500-square-meter high-tech workshop open to the public, which provides members with access to machines, tools and software as well as a creative community. The DIY machine shop and fabrication studio offers a place to implement ideas and innovations in the form of prototypes and small batch production. Various work areas are available, such as machine, metal and woodworking shops as well as textile and electrical processing facilities. In addition, 3D printers and laser and water jet cutters make it possible to fabricate new shapes and to process every type of material. MakerSpace offers training and consulting services as well as events for members with any level of knowledge, providing them with support and networking options. MakerSpace was designed and set up in cooperation with the Silicon Valley-based TechShop (see also Hatch, 2014).

A permanent challenge for start-up teams in the new venture creation journey is the constant lack of resources. To ease this problem, UnternehmerTUM serves as a unique platform for developing and connecting talent, technology, capital and customers. With its education and networking activities, UnternehmerTUM has built up an entrepreneurship community connecting more than 20,000 students, academics and professionals through online communities and communication channels as well as events such as 'Entrepreneurs' Night', when hundreds of young inventors, start-up teams and serial entrepreneurs meet in Munich. Most of these people are alumni of UnternehmerTUM courses. Networking activities are not only focused on a local level but also aim to strengthen international ties.

4.3 SYSTEMATIC SPIN-OFF PROCESS

The core competence of UnternehmerTUM is a systematic and efficient innovation and business creation process that turns ideas and technologies from science and industry into successful and sustainable products and services. UnternehmerTUM systematically identifies disruptive technologies and initiates new ventures. The entrepreneurship center focuses on high-technology start-ups in Information and Communication, Medical Engineering and CleanTech.

The goal of UnternehmerTUM is to generate more scalable high-tech start-ups. TUM and UnternehmerTUM actively support university start-up teams throughout all the development stages of a start-up, beginning with opportunity recognition and ending with the growth phase. This support includes technology scouting, opportunity assessment, team matching, consulting and coaching, customer acquisition and fundraising.

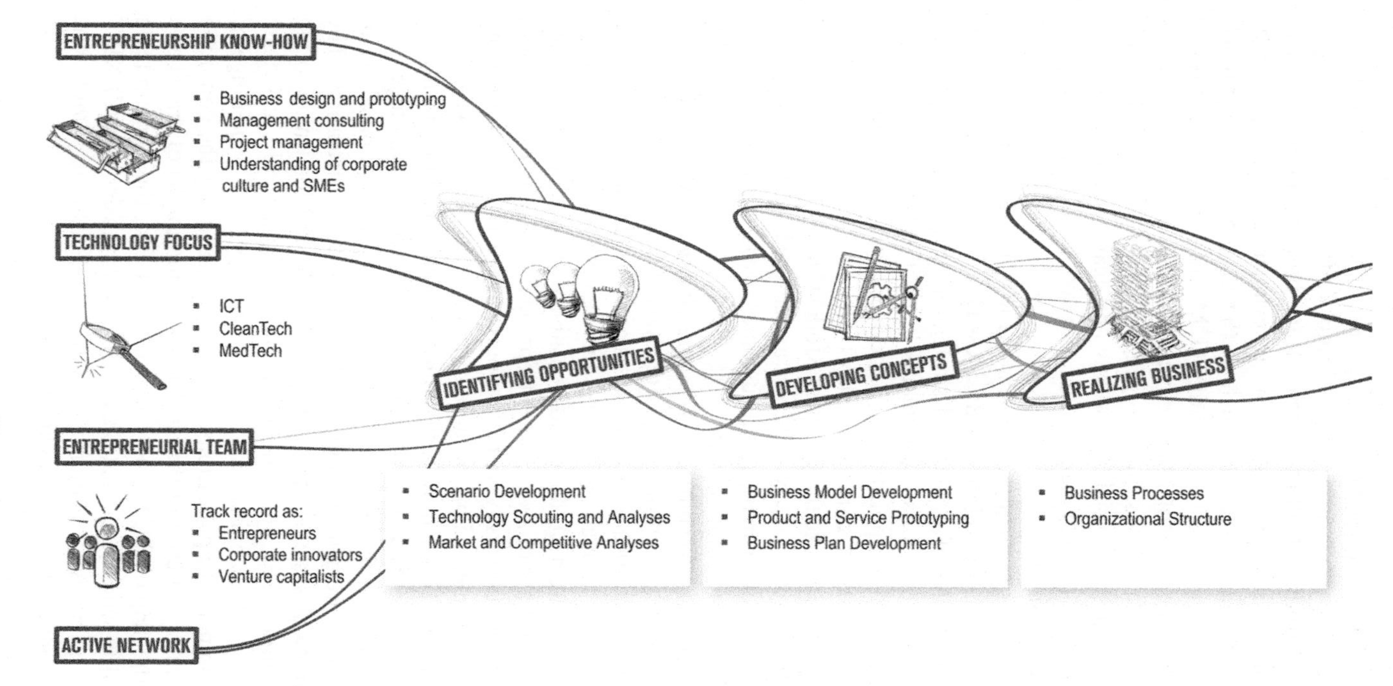
unternehmertum
Center for Innovation and Business Creation at TUM
ENTREPRENEURSHIP KNOW-HOW
Business design and prototyping
Management consulting
Project management
Understanding of corporate culture and SMEs
TECHNOLOGY FOCUS
ICT
CleanTech
MedTech
ENTREPRENEURIAL TEAM
Track record as:
Entrepreneurs
Corporate innovators
Venture capitalists
ACTIVE NETWORK
over 15,000 alumni
over 400 Startups
IDENTIFYING OPPORTUNITIES
DEVELOPING CONCEPTS
REALIZING BUSINESS
Scenario Development
Technology Scouting and Analyses
Market and Competitive Analyses
Business Model Development
Product and Service Prototyping
Business Plan Development
Business Processes
Organizational Structure

Figure 4.3 Core competence of UnternehmerTUM

Doctoral candidates can, for example, sign up for an intensive entrepreneurial qualification program called 'E-Camp! Enterprising Knowledge'. On this course, around 20 PhD students learn how to develop business ideas based on their research and how to evaluate market opportunities using state-of-the-art methods such as the Lean Start-up Approach (Blank, 2013). Experts from UnternehmerTUM, entrepreneurship professors from the TUM School of Management as well as successful founders inspire the students and assist them in generating creative ideas. In 2013, more than 200 entrepreneurial individuals and teams used the start-up service, leading to 47 new companies and more than ten business angel and venture capital investments.

Based on its long experience in hands-on start-up support, UnternehmerTUM has developed a 'Business Design Method Kit' assembling proven principles of design and strategic management as well as a set of best-practice tools for start-up teams. The kit distinguishes between three phases: 'identify opportunities', 'develop concepts' and 'realize business' and consists of cards with short descriptions of the method, examples and templates like the Business Model Canvas (Osterwalder and Pigneur, 2010). A broad variety of topics are covered, such as a comprehensive understanding of market and customer needs, business modeling, financial planning and team building.

One example of a TUM spin-off is *Dynamic Biosensors*, which has developed a patented biochip platform for the analysis of molecular interactions, enabling superior identification and analysis of pharmaceutical substances. Compared to conventional biosensors, the new system offers hitherto unobtainable measurement data and sensitivity. This globally oriented start-up took advantage of a series of entrepreneurship training and coaching activities at the university and is supported by a GO-Bio grant from the German government. For the development of its serial product and the upcoming market launch, *Dynamic Biosensors* raised additional money from UnternehmerTUM-Fonds and other venture capital companies.

Another start-up is *fos4X*, which develops and markets optical strain sensors and measurement technology for lightweight structures. The strain measurement systems of *fos4X* improve physical load monitoring of carbon composite or glass fiber structures, significantly increasing their efficiency and reliability. This novel technology has a wide range of interesting commercial applications, including monitoring the condition of wind generator blades and the electric-rod pantographs of trains and trams. With the help of TUM and UnternehmerTUM, the start-up team was successful in winning the national technology transfer grant 'EXIST Forschungstransfer'. After acquiring the first pilot customers, *fos4x* was also able to conclude two venture capital rounds.

4.4 INCUBATION PROGRAM KICKSTART

Inspired by US incubators and accelerators like Y Combinator, TechStars (see also The Power of Accelerators (Feld, 2012, p. 109), 500 Startups and StartX, UnternehmerTUM has, since, bundled its support for early-stage high-potential start-up teams in an incubation program called KICKSTART. The goal of this 12-month program is to successfully build outstanding start-ups and help participating teams to develop their entrepreneurial skills. The selected start-ups receive coaching, mentoring, access to office space and a prototyping infrastructure, as well as guidance on fundraising.

In the following paragraphs, the program is described in more detail according to the five main features of accelerators defined by Paul Miller and Kirsten Bound (Miller and Bound, 2011, p. 3). This is done to illustrate the similarities as well as the differences between KICKSTART and established accelerator models.

(1) *An application process that is open to all, yet highly competitive* KICKSTART has a highly selective admission process. Teams either have to be directly affiliated with TUM or already linked with UnternehmerTUM activities, e.g. in Climate-KIC of the European Institute of Innovation and Technology. Most applications come from start-ups in the Munich ecosystem. Pre-selected teams are invited to pitch in front of a jury. Important selection criteria are team commitment and know-how, innovativeness of technology, product, process or service, customer value and scalability. Up to 20 teams are admitted each year.

(2) *Provision of pre-seed investment, usually in exchange for equity* KICKSTART does not directly provide capital. Instead, the program helps start-ups to win public grants like EXIST Gründerstipendium and pilot customers from industry. The advantage of this approach is that founders do not have to provide equity at an early stage. On the downside, the bureaucracy and long waiting times associated with government grants can become a disadvantage. The ultimate goal of the incubation program is that start-ups become investor-ready within a short timeframe. In order to win business angels and venture capital investors in a seed-funding round, start-ups have to present a compelling business story. To succeed, they must focus on product development, first market traction and team building. Two pitch events serve as key milestones. A half-time pitch is set up as a trial run. During these informal pitching sessions, mentors, industry experts and the KICKSTART founder community give early

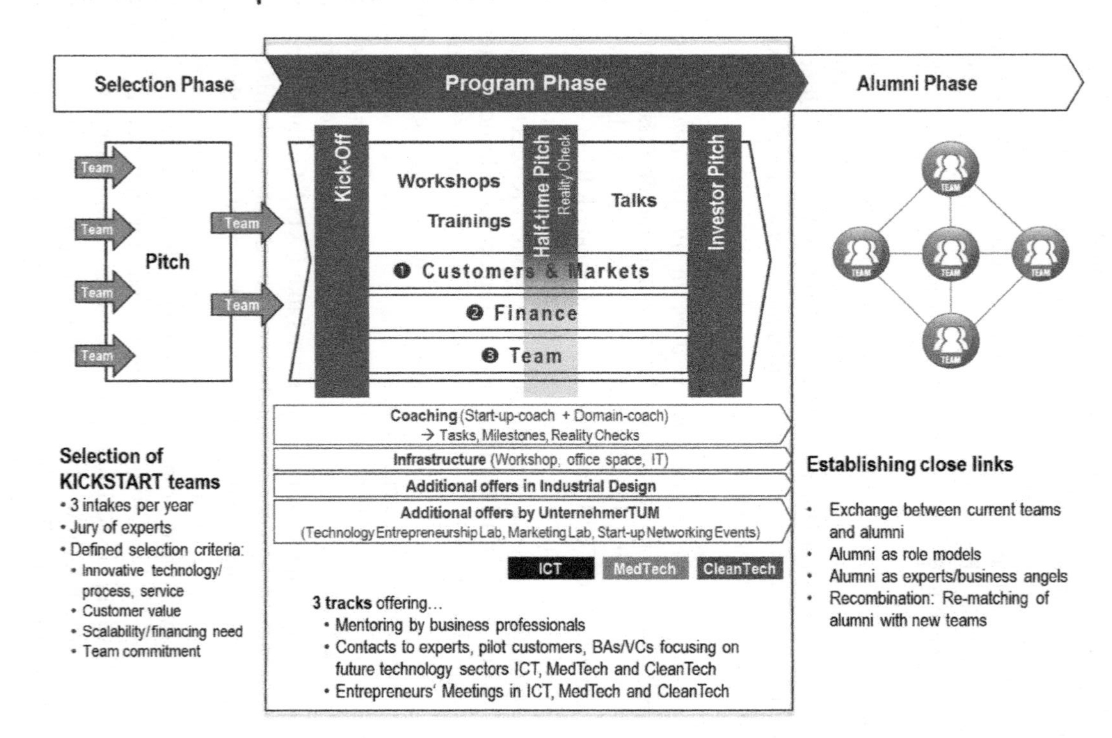

Figure 4.4 Incubation program KICKSTART

feedback. This includes feedback on the business model, the start-up strategy, the team and the presentation itself. The program culminates in a demo day. UnternehmerTUM brings business angels, venture capital investors and representatives from industry partners into direct contact with the KICKSTART teams and starts negotiations about investments and strategic partnerships.

(3) *A focus on small teams not individual founders* KICKSTART does not accept individual founders. During the application process and within the program, entrepreneurs are supported in team matching and team building activities. This includes the formation of an excellent, interdisciplinary and action-oriented core team as well as the recruiting and onboarding of the best employees available.

(4) *Time-limited support comprising programmed events and intensive mentoring* Via its extensive network, UnternehmerTUM is able to offer all KICKSTART teams access to experienced coaches, business experts, entrepreneurs and investors. Teams benefit enormously from the experience and feedback of these mentors. A series of workshops, training sessions and lectures also speeds up the learning process and the sharing of best practices.

(5) *Cohorts or 'classes' of start-ups rather than individual companies* KICKSTART has three intakes per year with five to ten teams per batch. By putting defined cohorts through the same process, start-ups learn with and from each other in workshops and networking events. This improves knowledge sharing and peer-to-peer learning. It also helps entrepreneurs to find synergies and to build a strong community of high-potential start-ups. This kind of intensive co-working can lead to a self-reinforcing entrepreneurial culture among the teams.

NavVis was one of the first teams accepted on the KICKSTART program. This technology company is a spin-off from the TUM Chair of Media Technology and was founded to create a better way of mapping, navigating and interacting virtually in indoor spaces. The computer vision-based technology, inspired by nature, is easy to use and does not require additional infrastructure. During KICKSTART, the start-up team was able to accelerate the development of its business model and successfully acquire pilot customers and a renowned angel investor by leveraging the UnternehmerTUM network.

4.5 COLLABORATION WITH ESTABLISHED COMPANIES

Thanks to Munich's strong industry cluster and the business-to-business focus of start-ups within the ecosystem, UnternehmerTUM is specialized in building business relationships between established companies and new ventures. Large firms can benefit enormously from the innovativeness and rapidity of entrepreneurs. Start-ups can derive immense advantage from winning an established company as a key customer, investor or strategic partner. By cooperating with corporates, new opportunities often evolve for start-ups, such as leveraging existing corporate sales channels for their own rapid international expansion.

UnternehmerTUM serves as a platform to connect and initiate collaborations (see also Schoenenberger, 2014). On the one hand, it helps established companies open up to start-ups. In consulting projects, corporate processes and structures are prepared to facilitate a cooperation project. It is often a big operational and cultural challenge for established companies to transform and be able and willing to buy from start-ups or even become strategic investors or partners. By introducing cooperation standards and even specialized collaboration programs, for example, company-supported accelerator programs like TechFounders and open innovation competitions, win-win partnerships between large companies and start-ups can be initiated quickly and easily. On the other hand, UnternehmerTUM delivers training and coaching for start-up teams to prepare them for interacting with corporates. Many founders lack understanding of and experience in dealing with corporate partners. Consequently, start-up acquisition activities to win investors and pilot customers are often doomed to failure. The training and workshops at UnternehmerTUM focus on topics such as customer or investor pitches, negotiating and contract management. Serial entrepreneurs, inventors, investors, potential pilot customers, industry experts and alumni are systematically involved in these sessions as guest speakers and mentors.

Market entry is a major task, especially for high-tech start-ups. UnternehmerTUM works with start-up teams to assess relevant market segments, target customers and position new businesses. With its broad network of established companies, the center acts as a door opener within the ecosystem. One example is Roding Automobile GmbH, a manufacturer of a two-seater mid-engine sports car with a carbon fiber chassis. The company was founded by former TUM engineering students who started building formula-style racing cars during their degree program. UnternehmerTUM supported the start-up team by providing the workshop infrastructure and access to established companies. Roding was able to win BMW as a partner and customer in the field of lightweight structure design. BMW is also providing a six-cylinder turbocharged engine for the sports car.

Since 2013, UnternehmerTUM has teamed up with Wissensfabrik, a non-profit association of German companies including BASF, Bosch, Fischer, ThyssenKrupp, TRUMPF and Voith, to strengthen cooperation models between corporations and young entrepreneurs throughout Germany. With more than 100 member companies, Wissensfabrik covers a wide variety of German industries. Together with media partner *Handelsblatt*, the leading German-language business newspaper, Wissensfabrik and UnternehmerTUM organize the annual German start-up competition WECONOMY. Following a rigorous selection process, around ten start-up teams are chosen by a high-profile jury and intensively coached. The highlight of the competition is a CEO weekend, when start-ups can discuss their business ideas one-to-one with top managers.

4.6 DISCUSSION OF CURRENT CHALLENGES AND THE FUTURE OUTLOOK

Since 2002, UnternehmerTUM has been continuously improving and enlarging its systematic innovation and business creation process. Nevertheless, TUM and its entrepreneurship center have yet to achieve the ultimate goal: to generate a significant number of new high-tech start-ups that evolve into global players. This could create a cluster dynamic with a self-reinforcing effect for more scalable start-ups.

UnternehmerTUM will therefore address several areas in the coming years. One major concern is that the spin-off potential at TUM is not yet fully realized. A critical mass of entrepreneurial-thinking researchers within the university's departments is still missing. Many more individuals within the university need to share and pursue the entrepreneurship mission. A growing output of successful founders and alumni who serve as role models will help to create a self-sustaining entrepreneurship culture in the university ecosystem.

The strategic approach to systematically cover the entire innovation and business creation process, starting with opportunity recognition and ending in the start-up growth phase, is very demanding. This 'one-stop-shop' approach leads to high operational complexity. To address all the critical issues within the enterprise creation process, a substantial number of experts, resources and know-how must be sourced. The success and sustainability of the innovation ecosystem depends heavily on long-term financial support and the backing of external sponsors, the university and state government. In a next optimization phase, UnternehmerTUM will deepen its know-how in special industry segments and related technologies to better support start-up projects in these fields.

For UnternehmerTUM, the most effective way to generate more scalable high-tech start-ups in the coming years is via its TechFounders accelerator program. To multiply the number of start-ups and speed up value creation, the entrepreneurship center will invest heavily in expanding its deal flow, improving its start-up support and attracting more experienced mentors, business angels and venture capitalists. Furthermore, UnternehmerTUM will strengthen its partnerships with established companies in TechFounders. Special tracks will focus on new market opportunities in areas such as the Internet of Things (IoT), automotive and 3D printing.

In the start-up/corporate collaboration context, the mobility of resources, especially people and capital, needs to be invigorated. To pursue large, scalable opportunities, entrepreneurs need access to sufficient risk capital. This can be provided by corporates and venture capitalists that singly or through syndication are able to invest several million euros in a single company. With adequate financial backing, scalable start-ups would also be more competitive in comparison with US-based, venture-capital-backed start-ups and could attract experienced professionals and entrepreneurs as co-founders and employees.

An additional important milestone was the opening in May 2015 of a new TUM and UnternehmerTUM entrepreneurship building on the research campus in Garching, just outside Munich. Dedicated to innovation and business creation, the 6,000-square-meter facility is designed to optimally host high-tech start-up teams by providing offices, space for training, networking and support services as well as the high-tech workshop infrastructure MakerSpace.

The new prototyping facility MakerSpace has been developed in cooperation with the BMW Group. This collaboration is designed to give business founders access to first-rate industry know-how. In return, BMW Group expects to be brought closer to groundbreaking innovations. Innovation initiated by BMW Group employees is also supported and the open workshop is available to them for prototyping activities.

UnternehmerTUM's unique offering is geared towards both German and international innovation and entrepreneurship, and aims to provide significant impetus for Bavaria's economy and innovative power. This should further reinforce Munich's position as a high-tech hub and strengthen its local network of universities, start-ups, companies and the creative scene.

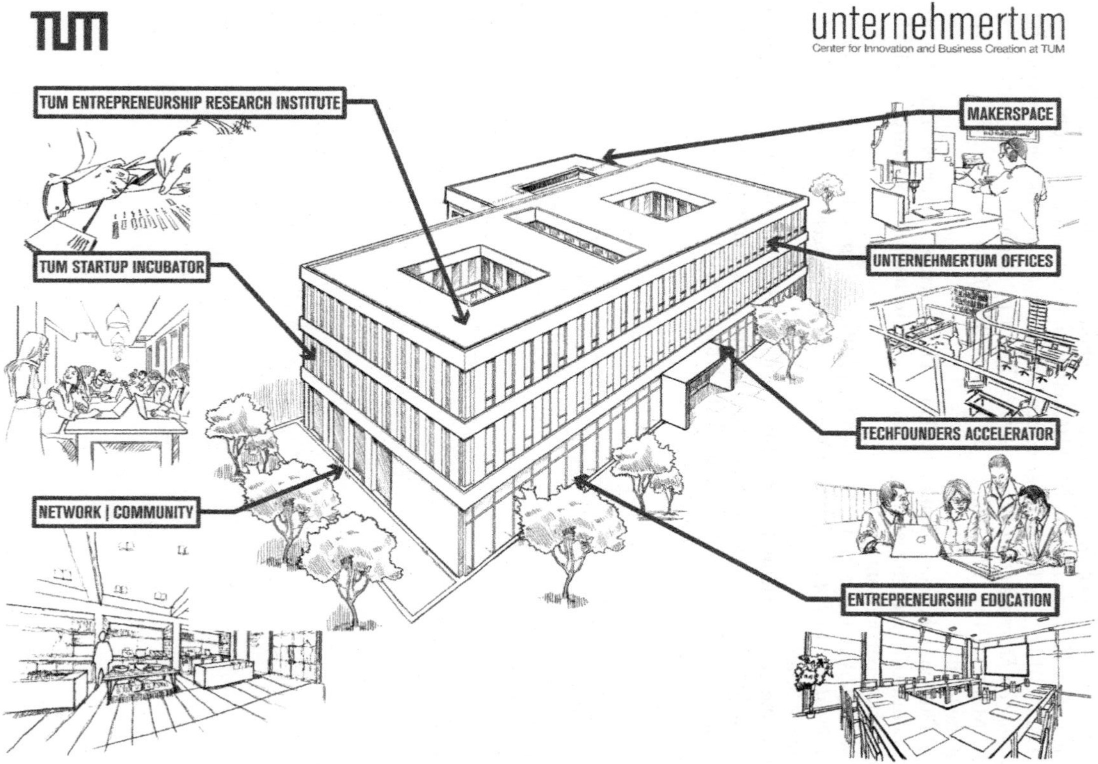

Figure 4.5 The new TUM and UnternehmerTUM entrepreneurship building

REFERENCES

Blank, S. (2013), 'Why the Lean Start-up Changes Everything', *Harvard Business Review*, May, 3–9.

Engel, Jerome S. (2014), *Global Clusters of Innovation, Entrepreneurial Engines of Economic Growth around the World* (Cheltenham: Edward Elgar Publishing).

Feld, Brad (2012), 'The Power of Accelerators', in Brad Feld, *Startup Communities, Building an Entrepreneurial Ecosystem in Your City* (Hoboken, NY: Wiley).

Hatch, Mark (2014), *The Maker Movement Manifesto* (New York City, NY: McGraw-Hill).

Miller, Paul and Kristen Bound (2011), 'The Startup Factories, The Rise of Accelerator Programmes to Support New Technology Ventures', NESTA Discussion paper, London.

Osterwalder, Alexander and Yves Pigneur (2010), *Business Model Generation: A Handbook for Visionaries, Game Changers, and Challengers* (Hoboken, NJ: John Wiley & Sons).

Schoenenberger, Helmut (2014), 'CollaborateToInnovate' (Munich: UnternehmerTUM, Wissensfabrik), https://www.unternehmertum.de/files/Handbuch_Collaborate.pdf (accessed 15 January 2016).

5. Supporting new spin-off ventures – experiences from a university start-up program

Magnus Klofsten and Erik Lundmark

5.1 INTRODUCTION

In the entrepreneurial society (Audretsch, 2009a, 2009b), universities have two important tasks: first to generate new knowledge and then to facilitate its practical application. To create and disseminate knowledge is by no means a new objective for universities. Two things have changed however – there is higher demand on the practical relevance of university research, and the effective implementation of new knowledge is no longer seen as the sole responsibility of government and large corporations. Rather, it is new, often small, knowledge-intensive companies that are seen as important actors in putting new knowledge into practice (Audretsch, 2009a, 2009b; Lundmark, 2010). Universities are increasingly expected to encourage and facilitate such organizations through spin-offs and incubator activities. Some claim that the universities themselves should be entrepreneurial (Clark, 1998). The entrepreneurial university as a concept is wide and includes many aspects (Gibb and Hannon, 2006; European Commission and OECD, 2013), such as partnerships between universities, businesses and the public sector (Etzkowitz and Klofsten, 2005; Brulin et al., 2012), focus on the application of research (Etzkowitz, 2001) and support for entrepreneurship among employees and students (Etzkowitz, 2004). This chapter deals with the latter, more specifically, how universities can facilitate the creation of new knowledge-intensive firms and new business areas within established organizations through practice-oriented entrepreneurship programs.

The chapter describes how the Entrepreneurship and New Business Program (ENP), which started at Linköping University in 1994, has evolved and spread to other parts of Sweden and Europe. The model underlying the program was developed at the Center for Innovation and Entrepreneurship (CIE) at Linköping University and the business network, Business Development in Linköping (SMIL).[1] Over the years, the ENP has attracted more than 1,500 participants and generated around 500 new businesses. The case illustrates how entrepreneurship training at universities with relatively simple measures can stimulate entrepreneurial action and learning.

The chapter begins with a discussion of what constitutes entrepreneurship and what implications this has for entrepreneurship training. Then we present an overview of the program, feedback from participants and a discussion of what constitute the key success factors in implementing entrepreneurship programs. The chapter concludes with a summary of the most important implications and lessons learned from many years of the ENP.

5.2 ENTREPRENEURSHIP AS ACTION AND BEHAVIOR

Entrepreneurship has been defined in many different ways, not only by academics but also by practitioners (Gartner, 1990). While some argue that entrepreneurship is about starting new organizations (Gartner, 1989; Aldrich and Ruef, 2006), others think that entrepreneurship is about innovation; that is, to successfully commercialize ideas (Schumpeter, 1934/2008; Drucker, 1985). A broader definition of entrepreneurship – the emergence of new economic activity (Davidsson and Wiklund, 2001; Wiklund et al., 2011) – has gained followers since the turn of the millennium. The latter and most recent definition of entrepreneurship thus includes commercial and non-commercial activity as well as business creation and development of existing businesses. What it boils down to is organizing new economic activities of some kind (note that 'economic' is wider than 'commercial'). Because the activities are new in some respects, it follows that uncertainty and unpredictability are central aspects of entrepreneurship.

Early entrepreneurship research, similar to early leadership research, focused on trying to identify what type of personality entrepreneurs have. Gartner (1989) argued against any attempt to distinguish entrepreneurs from non-entrepreneurs based on their personality. According to Gartner, decades of research had not generated any clear evidence that entrepreneurship can be identified through the personalities of the main actors in the process. Later research, through meta-studies, has established that certain personality traits actually correlate with the propensity to engage in entrepreneurship and with the level of success as an entrepreneur. However, there is consensus that the overlap between entrepreneurs and non-entrepreneurs, in terms of personality, is very large and that the average differences are small. Therefore, trying to predict who will become an entrepreneur, let alone who will be successful as an entrepreneur based on his/her personality, is not very fruitful (see Davidsson, 2013 for a concise research review).

Gartner (1989) argued that research should focus on behaviors rather than personality. From a training perspective, behaviors are also more promising because personality, by definition, is difficult to change. An important part of entrepreneurship training is thus focused on what activities and behaviors promote successful entrepreneurship. An important aspect of such training is to prepare (nascent) entrepreneurs in how to manage new and unpredictable situations (Sarasvathy, 2001).

5.3 THE DEVELOPMENT OF ENTREPRENEURSHIP PROGRAMS AT UNIVERSITIES

There are several obstacles to overcome in order to develop effective entrepreneurship programs at universities. Seen from the supply side, universities often lack the combination of theoretical and practical knowledge required to develop new business; furthermore, the programs that are implemented are in many cases not sufficiently long-term (Curran and Stanworth, 1989; Klofsten, 1994; Klofsten and Mikaelsson, 1996; Gibb, 2002). In addition, entrepreneurship as an academic discipline is relatively new and has not yet achieved the same legitimacy as more established fields (Vesper, 1988; Bygrave, 1994; Fayolle, 2013). Despite entrepreneurship being seen as something positive by the political establishment and society at large, some parts of academia still have negative attitudes towards entrepreneurship (Acs et al., 2009; Philpott et al., 2011; Lundmark and Westelius, 2014). From the demand side, it is common that prospective participants in an entrepreneurship program are skeptical about whether the program in question actually contributes relevant practical skills and whether participation is worth the time the program takes to complete (Gibb, 1990; Klofsten and Mikaelsson, 1996).

Garavan and O'Cinneide (1994) distinguish between different types of education and training activities aimed at entrepreneurs and small business owners. In particular, they demonstrate a conflict between management and entrepreneurship training; management training focuses on analysis and planning, while entrepreneurship training focuses on trial and error. However, participants in an entrepreneurship program often intend to start a business. Therefore, it is advisable to include classical aspects of management in the training program because the participants often need tools to analyze and manage markets and organizations (Klofsten, 1992).

While entrepreneurship programs have a connection to classical management training, and potential participants often see successful entrepreneurs as more credible than academics, it is not obvious who is best placed to lead entrepreneurship programs. Allowing a team consisting of both entrepreneurs and academics to lead entrepreneurial training programs has been highlighted as a successful model (Volkmann, 2004). Entrepreneurship training programs also differ from classical academic education. Rather than applying classical pedagogical approaches such as 'pulpit teaching', self-study and written examination, entrepreneurship programs tend to focus on practical exercises, case studies and working directly with the learners' ventures.

5.4 THE ENP: FOUNDING AND SUBSEQUENT DEVELOPMENT

The ENP was started at Linköping University (Center for Innovation and Entrepreneurship – CIE) in 1994. At the time, it was a step towards developing a more comprehensive offering of activities for supporting technology- and knowledge-intensive entrepreneurship. The university (and the SMIL network) had already accrued extensive experience in working with established firms such as Innovative Vision, Program System, Sectra and SoftLab. What was needed were activities for entrepreneurs in the start-up phase to, among other things, expand new business development at the university, improve the quality of new firms, and not least, if only indirectly, create an entrepreneurial university. The strategy of creating new spin-offs was not part of the university's strategy; the university administration neither actively supported the program nor hindered its growth. Rather, the program was the brainchild of a group of enthusiastic individuals from the university and SMIL.

Approximately ten people representing five potential firms participated in the first (pilot) program. The program was evaluated and received overwhelmingly positive feedback, which was incentive enough to continue. Interest in program participation continued to grow, not only in Linköping but also in other areas in Sweden and internationally. The program expanded from a fairly clearly defined concept – one of serving participants who intend to start up completely new firms in various sectors – to also include those who intended to start up new business units in established organizations.

Table 5.1 Examples of firms that have participated in the ENP program

Firm name	Year founded	No. of employees (2014)	Business sector	Spin-off background
Attentec	2005	30	Consulting firm in software development (mobility, internet and the web)	Epact Inc.
Applied Sensor (formerly Nordic Sensor Technologies)	1994	27	Chemical gas detector systems for appliances, cars, construction, and consumer and industrial applications	Linköping University: Department of Physics, Chemistry and Biology
Calluna	1992	35	Consultancy in nature conservation, on land and in water	Linköping University: Department of Physics, Chemistry and Biology
Dynamic Code	2002	8	Custom-designed package solutions for laboratory tests	The National Swedish Laboratory of Forensic Science and Board of Forensic Medicine
Idonex (formerly Informat-ionsvävarna & Roxen)	1994	35	Software for building and maintaining websites	Linköping University Department of Computer Science and Department of Physics, Chemistry and Biology
Kreatel (formerly Motorola, Google and now ARRIS)	1995	219	Service terminals for telecommunications and broadband networks	Linköping University: Department of Computer Science and Department of Physics, Chemistry and Biology
Wolfram Mathcore (formerly Mathcore)	1998	18	Mathematical modeling and simulation for effective product development processes	Linköping University: Department of Computer Science
Psykologpartners	2000	70	Consultancy based on cognitive behavioral therapy (CBT) and applied behavior analysis (ABA)	Linköping University: Department of Behavioral Science
Shapeline	1999	12	Systems for flatness measurements in the steel and metal industries	Epact Inc.
Xcerion (formerly Mediate)	2001	12	Holding company for innovations and patents in the field of internet-related technology	Linköping University: Department of Computer Science

The following is a summary of the development of the ENP in the past 20 or so years since its founding:

- Approximately 1,500 new entrepreneurs have participated in around 80 programs so far and have started some 500 firms and enterprises (see examples in Table 5.1). Outside the Linköping region, the program has been held in other Swedish regions – primarily Umeå (Uminova), Skara (Lyftet) and Västerås (Teknikbyn) – and internationally, in Russia (Obninsk) and Moldavia (Chişinău). Over half of all ENPs have taken place outside of the Linköping area.
- The ENP was one of the mainstays of UNISPIN (an EU project to develop university spin-off programs) to promote business development in a number of regions in Europe (van der Sijde and van Tilburg, 1998).
- Today, the program is offered in three forms: ENP *classic* (academic system), ENP *green industry* (agriculture, environment and energy), and ENP *organization* (public and private organizations).

Besides new enterprise creation, the program has been a source of numerous empirical studies that have led to articles in the *Journal of European Industrial Training* (Klofsten, 2000), *Regional Studies* (Klofsten et al., 2010), *Technology Analysis and Strategic Management* (Klofsten, 2005), and the *International Small Business Journal* (Davidsson et al., 2006).

5.5 CONTENT AND IMPLEMENTATION

Various skill sets and broad knowledge are needed to start a new business, which is why the entrepreneurship program offers a well-rounded package that includes activities for individual development (competence and driving force) and factors linked to firm identity (concept, product/service, market and organization) and to external firm relations (customer relations and other relations). Development in these areas hopefully leads to attainment of an idea platform – on which the concept is firmly grounded among relevant parties (Klofsten, 2005), and later perhaps even a business platform – on which the firm has achieved stable commercial operations (Klofsten, 1992; Davidsson and Klofsten, 2003).

An ENP comprises the following five cornerstones:

- *Business plan* or *plan of activities*. Each participant must develop a plan based on his/her own idea.

- *Workshops*. These focus on various aspects such as concept development, marketing, sales and funding.
- *Coaching*. Each program participant is scheduled for at least two meetings with a coach to discuss progress and solutions.
- *Mentorship*. Each participant is assigned a mentor who provides informal support and guidance on the entrepreneurial process.
- *Access to a network*. Each participant has access to the SMIL network as a guest, free of charge, during the program; this provides the opportunity to join various activities and meet other business owners and entrepreneurs.

The ENP differentiates between 'soft' and 'hard' resources. Soft resources refer to knowledge, relations, inspiration and other immaterial resources. Hard resources include money, infrastructure and other material resources. The ENP offers only soft resources. Previous research found that although both are important, they should not be mixed in the same program (Klofsten and Jones-Evans, 1996; Autio and Klofsten, 1998). The importance of involving participants in a networking process to effectively ground their concepts in their respective contexts cannot be emphasized enough (Olofsson, 1979). This is where not only SMIL's network of entrepreneurs contributes, but also investors, incubators and technology parks. Many network members participate in the program as workshop leaders, coaches and mentors. They also help recruit program participants. An important aspect of the program is its differentiation between coach and mentor. A coach works closely with the program management and has a central role in scheduling meetings to support and follow up work with business and operations plans; the role of a mentor is much more informal (Klofsten and Öberg, 2012). Both the coaches and mentors have extensive experience of starting and managing start-up businesses and they are all recruited from the SMIL network.

From the beginning, the program has allowed participants to work or study in parallel. Over time, however, the program has been shortened from one year to the current four to six months (see Table 5.2), which may limit the time for study or work. Shortening the program has increased its pace and participants' commitment.

5.6 TARGET GROUP AND RECRUITING

The ENP has several target groups, and these are primarily in the university environment (students, teachers and researchers), but some also come from firms and organizations (employees who start new activities in their organizations, or spin off their own company). The share of those

Table 5.2 A typical ENP schedule

Date/time	Contents
20 Aug/16:00–19:00	Kick-off, presentation of participants and other contributors. Introduction of coaching, homework and booking of meetings.
26 Aug/13:00–19:00	The Business Platform, cornerstones for the firm's early development. Group work.
11 Sept/13:00–18:00	Idea qualification/elimination – from idea to business idea.
Week 39	Coaching 1 – coaching in project groups.
30 Sept/13:00–19:00	Market and practical sales – marketing, how to sell, key customers and competitors.
9 Oct/13:00–19:00	Business planning and development plan – introduction to the business plan. Presentations of each group's project or development plan (5 min/project).
4th week Oct	Coaching 2 – coaching in project groups.
5 Nov/13:00–18:00	Leadership – project and group.
3rd week Nov	Coaching 3 – coaching in project groups.
26 Nov/13:00–19:00	Conclusion, project presentations (15 min/project) and award of certificates. A firm that has previously participated in the program presents itself. Feedback on individual presentations and the program as a whole.

from other backgrounds is relatively small. Recruiting criteria are simple – participants should have an idea (which does not need to be particularly well developed) but they must, above all, be strongly motivated to participate. Recruitment takes place via numerous channels: information flyers, newspaper announcements and advertising during other entrepreneurship courses. Perhaps the most important channel for reaching future participants is word of mouth, from program alumni to colleagues, and within the network of local support organizations.

Each participant is interviewed before being accepted to the program to ensure that the two criteria (concept and driving force) are fulfilled and to eliminate any misunderstandings that the ENP is a traditional, accredited course in entrepreneurship. There are two main reasons for using this relatively simple recruitment process.

(1) It is nearly impossible to determine whether an idea is good or bad at this stage. An early idea tends to change, and, in some cases, to change drastically during the maturation process (Timmons and Spinelli, 1994; Klofsten, 2005).

(2) Entrepreneurship is a process characterized by unpredictability and non-linearity, where the behavior of various individuals in interaction with their surroundings determines the end results (Sarasvathy, 2001; Lundmark and Westelius, 2012, 2014).

Thus, the program puts the individual (the entrepreneur or the entrepreneurial team) before the idea, which in practice means that if the idea is not robust, the idea is exchanged and not the individual. To develop an immature idea may require some time, especially if it has just emerged from the research environment, is abstract and is weakly grounded in the market (Klofsten, 2005).

One requirement, therefore, is full commitment from the participants, which includes taking advantage of the supporting people and the networks on offer during the program. The ENP can thus be considered an arena for activities and opportunities, ready to be used by participants who wish to develop their entrepreneurship.

5.7 PROGRAM FUNDING

Program participation is free of charge, and this decision was taken early on. The reason is that the typical participant is usually a young individual with no means of paying the fee required to cover the program's actual costs. Financing the program by selling shares has also been discussed, but it was considered unsuitable, primarily from an image standpoint. The program is not an 'investor activity' and it was considered important to have a neutral attitude to participants' activities, where all ideas have equal value. Instead, the program is publicly funded. The total cost of an average ENP is approximately SEK 350,000–500,000 (approximately US$54,500–78,000) for the financer. Normally, around 20 people with 10–15 ideas participate. Each idea is driven by one or more entrepreneurs. Each idea is assigned a mentor. At the start, participants usually pay a deposit of SEK 500 (approximately US$80) per group. This deposit is returned when the criteria for participation have been fulfilled. The aim is to ensure participants' intention to complete the program.

5.8 PARTICIPANT FEEDBACK

A simple evaluation is made at the end of each program, and the following positive aspects of participation have been cited:

- Development work became more structured.
- I was inspired to reach further – to accomplish more.
- The network we were introduced to was tremendously valuable.
- We were given a good look at what is actually required to start a firm and become a successful entrepreneur.
- I have become more professional.
- Various ways of looking at doing business.
- I feel much more sure of myself as an entrepreneur.
- I made a good start, and I might not have continued if it hadn't been for the program.

The most frequently cited positive feedback is access to the network through the program. The relationship with the mentor and the coach is also valued. And it is considered advantageous that participants are pushed to achieve more and structure work better. About one-fifth of participants do not finish the program for three main reasons: the original group was divided; the participant became pressed for time; or, simply, the idea was not feasible.

5.9 SUCCESS FACTORS IN ENTREPRENEURSHIP TRAINING

Based on approximately 20 years' experience of conducting the ENP, the following success factors must be highlighted:

- *Establish a comprehensive outlook* – Program content should cover a variety of business and operations development aspects.
- *Adapt the competence offering to the need* – Early development processes are dynamic and require that competences be effectively matched to the needs of participants. For example, it is important to match a group with suitable mentors and coaches who understand both the group and the idea. The mentors and coaches can then adapt program content to the group's needs.
- *Define real needs* – Participants are not always aware of their actual needs in the development process. Often, participants experience a need that should be analyzed before a relevant solution can be defined.
- *Use a network of complementary actors* – The program is more valuable if participants are given access to the activities and resources of other actors.
- *Promote participants' self-confidence* – At the beginning, participants usually lack practical experience of entrepreneurship, and they do

not consider themselves entrepreneurs. An important function of the program is to help participants feel inspired and comfortable in their new 'roles'.

- *Establish clear goals along the way* – Participants must be able to show that they are progressing from thought to action, which they document, for example, in the form of a simple business plan or in notes from meetings with potential customers. Participants must make a professional presentation of the business or operations plan at the conclusion of the program.
- *Use experienced entrepreneurs* – The backbone of the program is access to the experience and tested toolboxes of entrepreneurs who are willing to share their knowledge.
- *Put the right mentor together with the right participant* – The mentor and participant must match in factors such as personal chemistry, age, competence profile and mutual respect.
- *Sandwich internships with theory* – The program is practically oriented, but it is important to occasionally use research-based models that provide structure and clarity.
- *Recruit based on attitude, not background* – A mix of participants with various educational backgrounds and nationalities, of both genders, from different sectors, etc. makes interesting combinations of competence possible. It is important, however, to look at problems in new ways and promote a positive attitude to learning.
- *Engender trust* – Learning processes become easier in a climate of straightforward, open communication. Situations often arise that require confidentiality. However, confidentiality is difficult to manage relying only on contracts and legal obligations. The participants must feel that they can share their ideas and their knowledge when the entire group is assembled. So creating trust at an early stage is important; the open atmosphere to which this gives rise is necessary for interactive learning processes.
- *Avoid rushing results* – It takes time for people and ideas to develop. This should happen according to the ability and design in each individual case.
- *Strive for flexibility throughout the program* – Entrepreneurship is associated with action, new solutions to problems and new opportunities. This must be reflected in the program organization.

An important overriding point is that these success factors were compiled with a focal point on overall quality. It is important not to focus too much on how many firms were generated or how many participants chose not to pursue their ideas. What is important is that the program maintains a high level of quality and participants develop and acquire a better understanding of entrepreneurial processes. We believe that a focus on program quality – not start-up frequency – best serves successful entrepreneurship in the long run.

5.10 CONCLUSIONS AND IMPLICATIONS

In this chapter, we have focused on how a university can promote entrepreneurship through entrepreneurship training programs. We have argued that entrepreneurship is a behavior or a role that an individual or a group sometimes assumes rather than a personality trait or innate ability. The entrepreneurship program studied, ENP, shows that entrepreneurship can be encouraged and facilitated and that participating individuals and organizations often derive great benefit from attending. Sometimes this manifests itself in the creation of new businesses and sometimes it manifests itself through positive learning experiences of the participants (Fairlie et al., 2012).

The ENP has run for many years and has been tested in different environments and different countries. It is important to note that the program depends on support not only from external funding agencies but also from potential participants and internal stakeholders within the university. All three types of support, which Gibb (1990) would call 'gaps', must be in place for a functioning program. It is also important to note that it may take time to build an ENP. It took two to three years before the ENP really took off at the Linköping University. However, we have also found that when the program was transferred to other universities, it was implemented very quickly. In these cases, however, the program manager, the coach and the workshop leader had previous experience from Linköping University. When the host university itself has taken over management of the program, it has in some cases been very successful, while in others the host university has not been able to replicate the program due to, for example, lack of resources or the departure of a program champion.

Another important lesson is that the program contributes significantly to research at the university. An important implication of a successful entrepreneurship program is that many new organizations are created in close geographical and/or social proximity to the university. Thus the program makes it easier to get access to a large number of emerging businesses, which is central to entrepreneurship researchers. A less obvious but equally important aspect is that over the years many businesses grow, but maintain positive ties with the university and therefore benefit the university more broadly. For example, previous program participants, who may eventually develop successful businesses, constitute a source of mentors and networking partners for new participants. Because of its university ties, the ENP is a clear example of how research can be directly applied and utilized. In these ways, ENPs contribute to knowledge creation, as well as knowledge dissemination and application.

Although we believe that many of the success factors we describe above are universal, it is still important that each factor is assessed based on the context in which the program is implemented. The ENP has so far mainly been targeted at knowledge-intensive companies and run in university environments. It is thus important to highlight that the program must be adapted to the different backgrounds and objectives of participants. They are central because an important success factor of the program is to connect participants with each other and with mentors and trainers in order to generate value-creating networks.

NOTE

1. The network was initially called 'Stiftelsen för småföretagsutveckling' ('The foundation for small business development'), but the name changed as several members had grown so large that they could hardly be called small. However, the acronym SMIL was kept.

REFERENCES

Acs, Z.J., D.B. Audretsch and R.J. Strom (2009), 'Why entrepreneurship matters', in Z.J. Acs, D.B. Audretsch and R.J. Strom (eds), *Entrepreneurship, growth, and public policy* (New York, NY: Cambridge University Press), pp. 1–14.

Aldrich, H.E. and M. Ruef (2006), *Organizations evolving* (2nd edn) (Thousand Oaks, CA: Sage).

Audretsch, D.B. (2009a), 'The emergence of the entrepreneurial society', *Business Horizons*, **52** (5), 505–11.

Audretsch, D.B. (2009b), 'The entrepreneurial society', *Journal of Technology Transfer*, **34** (3), 245–54.

Autio, E. and M. Klofsten (1998), 'A comparative study of two European business incubators', *Journal of Small Business Management*, **36** (1), 30–43.

Brulin, G., P.-E. Ellström and L. Svensson (2012), 'Partssamverkan för effektiva produktionssystem och tillväxt', in L. Magnusson and J. Ottosson (eds), *Den hållbara svenska modellen: innovationskraft, förnyelse och effektivitet* (Stockholm: SNS förlag), pp. 69–89.

Bygrave, W.D. (1994), *The portable MBA in entrepreneurship* (Toronto, Ontario: John Wiley and Sons).

Clark, B.R. (1998), *Creating entrepreneurial universities: organizational pathways of transformation* (Issues in Higher Education) (New York, NY: Elsevier Science Regional Sales).

Curran, J. and J. Stanworth (1989), 'Education and training for enterprise:

problems of classification, evaluation, policy and research', *International Small Business Journal*, **7** (2), 11–22.
Davidsson, P. (2013), *Is entrepreneurship a matter of personality?* ACE Research Vignette, http://eprints.qut.edu.au/63369/ (accessed 3 May 2016).
Davidsson, P. and M. Klofsten (2003), 'The business platform: Developing an instrument to gauge and to assist the development of young firms', *Journal of Small Business Management*, **41** (1), 1–26.
Davidsson, P. and J. Wiklund (2001), 'Levels of analysis in entrepreneurship research: Current research practice and suggestions for the future', *Entrepreneurship Theory and Practice*, **25** (4), 81–99.
Davidsson, P., E. Hunter and M. Klofsten (2006), 'Institutional forces: The invisible hand that shapes venture ideas?', *International Small Business Journal*, **24** (2), 115–31.
Drucker, P.F. (1985), *Innovation and entrepreneurship* (London: Heinemann).
Etzkowitz, H. (2001), 'The second academic revolution and the rise of entrepreneurial science', *Technology and Society Magazine, IEEE*, **20** (2), 18–29.
Etzkowitz, H. (2004), 'The evolution of the entrepreneurial university', *International Journal of Technology and Globalisation*, **1** (1), 64–77.
Etzkowitz, H. and M. Klofsten (2005), 'The innovative region: Toward a theory of knowledge-based regional development', *R&D Management*, **35** (3), 243–55.
European Commission and OECD (2013), *A guiding framework for entrepreneurial universities*, Internal report, Brussels.
Fairlie, R.W., D. Karlan and J. Zinman (2012), *Behind the gate experiment: Evidence on effects of and rationales for subsidized entrepreneurship training* (National Bureau of Economic Research), http://dl.kli.re.kr/dl_image/IMG/03/000000010896/SERVICE/000000010896_01.PDF (accessed 3 May 2016).
Fayolle, A. (2013), 'Personal views on the future of entrepreneurship education', *Entrepreneurship & Regional Development*, **25** (7–8), 692–701.
Garavan, T.N. and B. O'Cinneide (1994), 'Entrepreneurship education and training programmes: A review and evaluation – part 1', *Journal of European Industrial Training*, **18** (8), 3–12.
Gartner, W.B. (1989), 'Who is an entrepreneur? Is the wrong question', *Entrepreneurship Theory and Practice*, **13** (4), 47–68.
Gartner, W.B. (1990), 'What are we talking about when we talk about entrepreneurship?', *Journal of Business Venturing*, **5** (1), 15–28.
Gibb, A. (1987), 'Designing effective programmes for encouraging the business start-up process: lessons from UK experience', *Journal of European Industrial Training*, **11** (4), 24–32.
Gibb, A. (1990), 'Design effective programmes for encouraging the small business start-up process', *Journal of European Industrial Training*, **14** (1), 17–25.
Gibb, A. (2002), 'In pursuit of a new "enterprise" and "entrepreneurship" paradigm for learning: Creative destruction, new values, new ways of doing things and new combinations of knowledge', *International Journal of Management Reviews*, **4** (3), 233–69.
Gibb, A. and P. Hannon (2006), 'Towards the entrepreneurial university?', *International Journal of Entrepreneurship Education*, **4**, 73–110.
Klofsten, M. (1992), 'Tidiga utvecklingsprocesser i teknikbaserade företag', Thesis, Linköping University.
Klofsten, M. (1994), 'Technology-based firms: critical aspects of their early development', *Journal of Enterprising Culture*, **2** (1), 535–57.

Klofsten, M. (2000), 'Training entrepreneurship at universities: A Swedish case', *Journal of European Industrial Training*, **24** (6), 337–44.
Klofsten, M. (2005), 'New venture ideas: An analysis of their origin and early development', *Technology Analysis and Strategic Management*, **17** (1), 105–19.
Klofsten, M. and D. Jones-Evans (1996), 'Stimulation of technology-based small firms: A case study of university-industry cooperation', *Technovation*, **16** (4), 187–213.
Klofsten, M. and A.S. Mikaelsson (1996), 'Support of small business firms: Entrepreneurs' views of the demand and supply side', *Journal of Enterprising Culture*, **4** (4), 417–32.
Klofsten, M. and S. Öberg (2012), 'Coaching versus mentoring: Are there any differences?', *New Technology-Based Firms in the New Millenium*, **9**, 39–47.
Klofsten, M., P. Heydebreck and D. Jones-Evans (2010), 'Transferring good practice beyond organizational borders: Lessons from transferring an entrepreneurship programme', *Regional Studies*, **44** (6), 791–9.
Lundmark, E. (2010), *The mobility of people, ideas and knowledge in the entrepreneurial society*, Thesis, Linköpings University, http://liu.diva-portal.org/smash/record.jsf?pid=diva2%3A380490&dswid=5934.
Lundmark, E. and A. Westelius (2012), 'Exploring entrepreneurship as misbehavior', in A. Barnes and L. Taksa (ed.), *Rethinking misbehavior and resistance in organizations* (*Advances in industrial & labor relations, Volume 19*) (Bingely: Emerald Group Publishing Limited), pp. 209–35.
Lundmark, E. and A. Westelius (2014), 'Entrepreneurship as elixir and mutagen', *Entrepreneurship Theory and Practice*, **38** (3), 575–600.
Olofsson, C. (1979), *Företagets exploatering av sina marknadsrelationer. En studie av produktutveckling*, Forskningsrapport Nr. 91, Ekonomiska Institutionen, Linköping University.
Philpott, K., L. Dooley, C. O'Reilly and G. Lupton (2011), 'The entrepreneurial university: Examining the underlying academic tensions', *Technovation*, **31** (4), 161–70.
Sarasvathy, S.D. (2001), 'Causation and effectuation: Toward a theoretical shift from economic inevitability to entrepreneurial contingency', *Academy of Management Review*, **26** (2), 243–63.
Schumpeter, J.A. (1934/2008), *The theory of economic development: An inquiry into profits, capital, credit, interest and the business cycle* (London: Transaction Publishers).
Timmons, J.A. and S. Spinelli (1994), *New venture creation: Entrepreneurship for the 21st century* (Boston: Irwin), vol. 4.
van der Sijde, P. and J. van Tilburg (1998), 'Creating a climate for university spin-offs', *Industry and Higher Education*, **12** (5), 297–302.
Vesper, K.H. (1988), 'Entrepreneurial academics: How can we tell when the field is getting somewhere?', *Journal of Business Venturing*, **3** (1), 1–10.
Volkmann, C. (2004), 'Entrepreneurial studies in higher education: Entrepreneurship studies: an ascending academic discipline in the twenty-first century', *Higher Education in Europe*, **29** (2), 177–85.
Wiklund, J., P. Davidsson, D.B. Audretsch and C. Karlsson (2011), 'The future of entrepreneurship research', *Entrepreneurship Theory and Practice*, **35** (1), 1–9.

6. 'Intrapreneurship at the Fraunhofer-Gesellschaft': how to stimulate greater entrepreneurship among researchers

Julia Bauer, Matthias Keckl, Thorsten Lambertus and Björn Schmalfuß

6.1 INTRODUCTION

The Fraunhofer-Gesellschaft is Europe's largest application-oriented research organization, carrying out practical research in close cooperation with its customers. At present, Fraunhofer maintains 67 institutes and research units across Germany and also operates subsidiaries abroad.

Fraunhofer Venture is the technology transfer office for Fraunhofer start-ups. However, the number of spin-offs emerging from Fraunhofer institutes is stagnating. This is due to Fraunhofer's strong links to industry as well as a flourishing economy. Due to the high number of industry projects, Fraunhofer researchers have limited time to work on their own ideas; furthermore, in the current economy, the industry offers well-paid and stable jobs. Moreover, the senior management of Fraunhofer institutes is hesitant to support spin-offs because talented researchers with technological know-how could then leave their institutes.

The aim of the Fraunhofer Venture Lab initiative is to nurture a sustainable intrapreneurial mindset within Fraunhofer in order to stimulate entrepreneurial behavior and eventually increase the number of spin-offs.

To achieve this goal, the Fraunhofer Venture Lab sets up intrapreneurship approaches at Fraunhofer institutes based on three pillars:

(1) *Sensitization*: to make Fraunhofer researchers aware of different career options by introducing them to role models such as experienced Fraunhofer intrapreneurs and entrepreneurs.
(2) *Ideation*: translating a technology into a business idea.

(3) *Business Design*: a systematic approach to developing a business idea into a business model.

The Venture Lab also offers Fraunhofer intrapreneurs with a valid business model the opportunity to take part in so-called Fraunhofer Days or 'FDays': a 12-week internal accelerator program providing intensive coaching and time to further develop their business model.

In sum, the Venture Lab initiative aims to nurture more innovation and new business ideas within Fraunhofer institutes and to increase the number of spin-offs initiated by intrapreneurial employees.

6.2 THE FRAUNHOFER MODEL AND THE CHALLENGES OF NURTURING AN ENTREPRENEURIAL CULTURE

6.2.1 The Structure of the Fraunhofer-Gesellschaft: A Specific Approach to Supporting SMEs through R&D

At present, the Fraunhofer-Gesellschaft maintains 67 institutes and research units across Germany and also operates subsidiaries abroad. Compared to other publicly funded R&D organizations and universities, the Fraunhofer model is unique in cooperating with industry partners. As the largest organization for applied research in Europe, Fraunhofer carries out practical research in close cooperation with its customers, which range from large corporations to small and medium-sized enterprises (SMEs). This approach helps to shape the innovation process in Germany and stimulates the development of future key technologies. The majority of Fraunhofer's almost 24,000 staff members are qualified scientists and engineers. Fraunhofer institutes are located all over Germany and have different technical competences. Therefore, there is a great diversity of innovation cultures within the organization.

Fraunhofer's annual research budget reached 2.1 billion euros in 2015. Of this, 1.8 billion euros was generated through contract research. In general, more than 70% of the contract research revenue is derived from contracts with industry partners and publicly financed research projects. The German federal and state governments contribute almost 30% in the form of base funding.

The most important way of transferring technologies to the market is cooperating with and working for industry partners. In 2015, Fraunhofer institutes acquired more than 600 million euros from industry partners. However, over the years the importance of other ways of technology

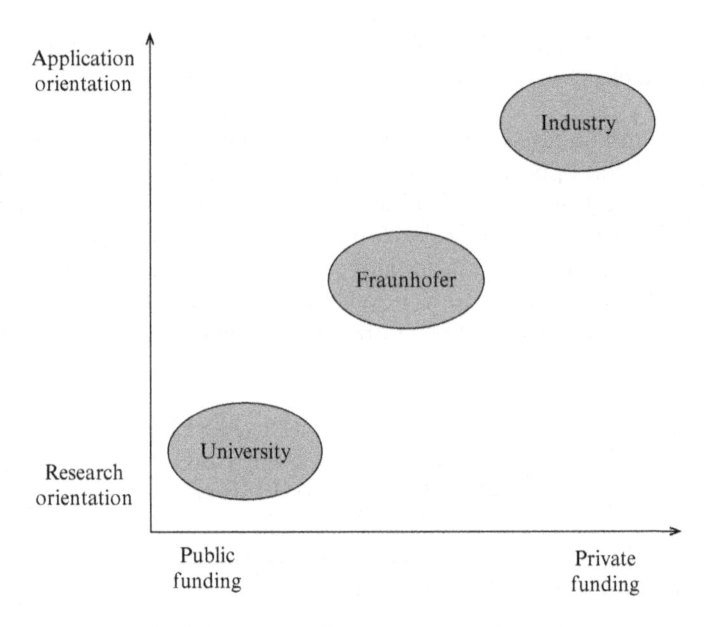

Figure 6.1 Fraunhofer's position between university and research

transfer has grown dramatically, and, in particular, the success in licensing MP3 technology has proved to be a useful strategy for generating new revenues.

Fraunhofer institutes operate in a multi-dimensional field of tensions. Their customers and public partners expect excellent research combined with a clear link to relevant market needs. This presents an enormous challenge for managing a Fraunhofer unit and also for the research staff. To continue the success of the Fraunhofer Society, researchers and administrations are forced to implement new entrepreneurial approaches that combine excellent research with market-relevant solutions.

6.2.2 Fraunhofer Institutes: Decentralized and Embedded in Strong Local Ecosystems

The decentralized structure of the Fraunhofer-Gesellschaft is based on the idea of achieving close proximity to industry and universities in order to create a bridge between basic research and industry applications. Therefore, each institute's director holds a chair at a university close by. This gives students the opportunity to connect with Fraunhofer and experience applied research processes early in their studies, e.g. as research

Figure 6.2 Fraunhofer locations across Germany

assistants. Students then often stay at Fraunhofer during their doctoral studies to combine their academic research with a business-oriented contract research approach, which includes understanding customer needs, the fulfillment of contract obligations, the acquisition of new customers and budgets. As a result of working in close proximity to industry partners, many employees leave Fraunhofer to start their careers in these companies. However, they remain in touch with Fraunhofer for special research tasks that they know Fraunhofer can deliver. This creates an ecosystem in which former employees become ambassadors for Fraunhofer, providing new requests and research challenges from industry.

Close collaboration with universities also embeds Fraunhofer in the regional, national and international scientific community. Together with universities and other extramural research organizations such as the Helmholtz Association, Gottfried Wilhelm Leibniz Association and the Max-Planck-Society, Fraunhofer contributes to Germany's world-class research landscape. This includes basic research funded by the German government or the European Union and applied research settings where Fraunhofer works in close collaboration with local medium-sized companies as well as large corporations and universities to develop solutions in a variety of technological domains.

For example, the Fraunhofer institute for production and automation (IPA) and the Fraunhofer institute for industrial engineering (IAO) in Stuttgart have joined forces with institutes at Stuttgart University to develop one applied research center on the topic of 'Industry 4.0'. Many medium-sized companies in the region, but also large corporations like Hewlett-Packard and Siemens, seek Fraunhofer's support to prepare for the highly automated and connected industrial production environment of the future. This center is contributing to Europe's leading region for automation research and work science, where every third job is already related to these topics.

However, Fraunhofer institutes are not only linked to universities; they are also interconnected via industry-specific 'Fraunhofer Groups' and topic-specific 'Fraunhofer Alliances'. On the one hand, this leads to a pooling of different kinds of expertise within Fraunhofer, an opportunity to create system solutions and facilitate access for customers to services in specific fields of technology. On the other hand, because the institutes have a high level of independence through the profit-center-based structure of the Fraunhofer-Gesellschaft, they compete with one another to gain the best industry contracts.

6.2.3 Impact of Industry Orientation of Fraunhofer Institutes on Spin-Off Activities

Fraunhofer is the outsourced R&D lab of Germany's well-known 'Mittelstand' category (small and medium-sized companies). As already mentioned, more than 70% of Fraunhofer's research revenue is derived from contracts with industry and publicly financed research projects. Tying a significant amount of Fraunhofer's funding to collaborative projects with companies forces the institutes to focus their research activities on what is currently needed on the market. Every euro that results from industry projects is rewarded with a bonus from basic financing. This incentive ultimately leads to the very strong industry orientation of all institutes. With regard to spin-offs, this macroeconomic benefit for Germany's Mittelstand comes at a price: the huge potential to bring technologies to market via spin-offs is not leveraged. There are two reasons for this:

(1) *Researchers*: industry projects are time-consuming. Also, developing vital spin-off opportunities usually takes time. The resources needed for project acquisition, project management and technology development limit the time for entrepreneurial activities considerably. Nevertheless, some researchers identify business opportunities and often gain insights from industry projects. However, especially in a thriving economy – and in recent years, the German economy has been developing favorably – companies pay higher salaries. In combination with the relatively risk-averse German culture, the idea of founding a company is less appealing to many researchers than life as a well-paid employee at one of Germany's strong companies such as BMW, Siemens, Bosch or Bayer, to name but a few. Thus, most Fraunhofer researchers strive for an industry career.
(2) *Fraunhofer institutes*: the numbers from Fraunhofer Venture's portfolio companies (currently >80 Fraunhofer spin-offs) clearly show that spin-offs in general, and especially successful ones, account for a considerable amount of industry money flowing back to the institutes from where they were started. Yet, each institute must strategically decide how to exploit its technologies. Most of them fear licensing their best technologies to spin-offs. If a technology is exclusively used by an unsuccessful spin-off, then another firm cannot use it at the same time. As a consequence, most institutes decide not to bet on spin-offs as a carrier of their technologies to market. Further aspects are: (a) most institutes choose short-term results over long-term spin-off exit opportunities; and (b) not only do the best technologies leave the institute, but often also their best human capital.

Due to these facts, a spin-off is a more complex decision at Fraunhofer compared to common industry projects. If there is no additional incentive, Fraunhofer institutes are prone to eschew or delay entrepreneurial activities.

6.3 INTRAPRENEURSHIP AT THE FRAUNHOFER-GESELLSCHAFT

6.3.1 Intrapreneurship as a Solution: Stimulating More Entrepreneurship by Technological Competence Owners within Fraunhofer

We can see from other examples in this book that technology transfer via spin-offs has numerous positive effects on the company or, in our case, the institute from where the technology originates. However, as explained above, due to the strong industry focus of Fraunhofer institutes, spin-offs are generally not the preferred way to transfer a technology to the market.

One striking difference between technology transfer via a spin-off company and other forms of technology transfer (e.g. licensing) is that a founding team with specific skills and an entrepreneurial mindset is needed to successfully transfer a technology to the market. However, most of the time, and due to their day-to-day industry focus, researchers do not have strong entrepreneurial ambitions or mindsets. Hence, they are less likely to consider career paths outside industry or research.

Thus, we see an early and sustainable implementation of intrapreneurial skills and attitude within Fraunhofer institutes as a potential means to overcome this challenge.

What do we mean by intrapreneurial skills and attitude? The term 'intrapreneur' is defined as a person acting like an entrepreneur, but as an employee within the boundaries of the firm. Such an intrapreneur proactively drives innovation. He/she recognizes, seizes and exploits opportunities proactively. The term 'intrapreneurship' can be traced back to Pinchot and Pinchot (1978). Several firms, including 3M, Google, Sony, Facebook and Sun Microsystems, have successfully implemented intrapreneurship approaches within their companies since then.

Why do we focus on intrapreneurship? An intrapreneurship – sometimes also labeled corporate entrepreneurship – program focuses on people. Well-educated, motivated and intrapreneurial people especially are the drivers for innovation and creativity (Martins and Terblanche, 2013). The more people think in an intrapreneurial and proactive way, the more innovative ideas appear and enter the innovation funnel. If a larger number of elaborated projects driven by intrapreneurial people enter the funnel, a larger number of innovative business ideas will result.

Because of that, we, the Venture Lab team and authors of this chapter, want to nurture intrapreneurship basics within each Fraunhofer institute. Our aim is to improve the intrapreneurial mindset within the Fraunhofer-Gesellschaft and, as a consequence, also the entrepreneurial mindset, resulting in a larger number of spin-offs.

The basics of intrapreneurship require equipping staff members with tools and methods from the start-up world (e.g. the Lean Start-up method) that are also applicable to innovation projects in their early stages. For example, our aim is to make technological competence owners aware from early on that technology development has to be aligned with market needs and customer requirements. All too often, technological competence owners develop technologies over a long period of time before looking for a market and then realizing that the market needs something different, or that the technology is far too complex and has too many features that customers do not require. Thus, learning about market needs and customer requirements beforehand as well as testing and validating technological prototypes on the market at an early stage is essential. In this way, technological competence owners can learn fast and, in some cases, fail fast. Every intrapreneur needs this attitude.

Further, we want to open up new career opportunities for researchers. When researchers are equipped with intrapreneurial skills, the step towards becoming entrepreneurs comes closer. Intrapreneurship also has great benefits for the institute itself. Intrapreneurial researchers are better at recognizing opportunities, such as identifying new business areas or creating new licensing models for the institute.

In summary, the aim of our intrapreneurship approach is to foster intrapreneurial thinking among staff members early on, equipping them with the tools and methodologies they need to transfer technologies to the market and stimulating an intra- and entrepreneurial culture within Fraunhofer institutes.

6.3.2 Intrapreneurship at Fraunhofer: Focus on People instead of Technology Projects

A fundamental challenge for every organization is transforming initiatives such as the Venture Lab into self-sustaining processes. This, of course, also strongly affects how we think about implementing intrapreneurship at the Fraunhofer-Gesellschaft in a sustainable way. First, as representatives of Fraunhofer HQ we cannot force, but need to convince every single institute to adopt our intrapreneurship approach. Secondly, the most important building blocks of intrapreneurship must eventually be deeply embedded in the organization as a whole and thus also in the institutes. That is why we follow a capacity-building approach. The entrepreneurial-minded individual propels innovation and thus has to be:

- empowered with tools and methodologies;
- provided with necessary resources;

- protected from obstacles that might block their path;
- connected with complementary networks.

If one is able to implement a fully functioning entrepreneurship ecosystem in an institute, the chances are high that intrapreneurship and also entrepreneurship will flourish. In order to reach that goal, we are offering to build Intrapreneurship Labs. An Intrapreneurship Lab brings together all the necessary elements for entrepreneurship, which means that it empowers, provides resources, protects and connects.

We partner with pilot institutes in order to test and learn about the necessary feature sets of such Intrapreneurship Labs. We have deliberately chosen a heterogeneous sample of institutes, which allows us to draw insights from different cultures, organizational structures and processes found at different Fraunhofer units. What has already become clear is that there is no 'one-size-fits-all' solution. The commitment of the institutes is essential for arriving at a sustainable solution.

In the following paragraphs we outline the fundamental pillars of the current version of our intrapreneurship approach.

6.3.2.1 Awareness and sensitization

The first pillar acts as the basis for and support of the following pillars. It makes researchers aware of intrapreneurial thinking and acting, and sparks their interest in the topic. Here, role models from Fraunhofer (intra- and entrepreneurs) play an important role. These role models are people who were once in the same position as current Fraunhofer researchers and with whom researchers can identify. They introduce new career perspectives and possibilities to researchers. Role models are either presented online in videos or invited to offline events, such as talks or other entrepreneurial events. Furthermore, we offer online training in intra- and entrepreneurship tools and skills that are relevant for both intrapreneurs and entrepreneurs.

6.3.2.2 Ideation and idea platform

The aim of the ideation pillar is a conjunction of technological solution and market problem. In this phase, a first business idea is derived from Fraunhofer technologies. Although Fraunhofer technologies are mostly invented for and together with our business partners, there is a major gap between technology and market that we try to bridge in the ideation phase.

This can be done in two ways. The most common way (and an essential part of our ideation pillar) is to look for market problems that fit Fraunhofer's technological solutions. However, we also ideate around actual market problems of industry partners and seek solutions within our technological portfolio.

Different tools are available to support proactive, intrapreneurial researchers who have a technology idea that might potentially be translated into a business idea. In different workshop formats, like the technological competence leveraging workshop (a systematic approach to identifying future fields of application for a technology) or the idea workshop or the idea game, researchers are supported to answer the following questions:

- What is a profitable field of application for my technology?
- Who are the customers and who are the users?
- What product or service can we offer our customers?
- What are the jobs our customers try to get done? What are their pains and gains?
- What value can we generate for our customers/users with our offer?

In these workshops we use elements from Technological Competence Leveraging (Keinz and Prügl, 2010), The Lean Start-up (Blank, 2013; Blank and Dorf, 2012; Ries, 2011), Value Proposition Design (Osterwalder et al., 2014) and Design Thinking (Brown, 2008; Kelley, 2001; Wylant, 2008), but adapt them to the specifics of a large application-oriented research organization like Fraunhofer. Additionally, we add specific skills training for our intrapreneurs based on those methods, such as communication skills (pitching, storytelling, prototyping, etc.) and market research skills (qualitative customer interviews, customer observation, etc.).

Furthermore, to support Fraunhofer intrapreneurs on their journey from an initial idea to a viable business model, we have developed an online idea platform to back the different real-life offerings (the 'Fraunhofer Ideenportal'). Here, users can share new ideas, discuss them with our project team and other experts, validate their business models and document results as well as improvements along the way. Through templates and additional materials, we further help intrapreneurs to improve their skills and interconnect along the different stages of the process.

To increase the number of Fraunhofer employees we reach, a Fraunhofer-wide idea competition was launched in collaboration with another HQ department, and participants were invited to attend specific workshops and events subsequently.

6.3.2.3 Business Design

Business Design is our fundamental approach to bringing business ideas to market and, at the same time, is one of the major stages of Fraunhofer's intrapreneurship activities. The term 'Business Design' already highlights two important characteristics: first, it is pure business. Without neglecting the fact that we operate in a high-tech environment, the focus is on a deep understanding of customers, markets, business models and business eco-systems. Secondly, we are not just analysing business cases; we proactively exploit situations, come up with creative ways to do business and draw upon the iterative nature of designers' innovation processes. Like the Ideation pillar, Business Design combines and enhances the key principles of modern innovation methodologies such as the Lean Start-up, Business Model Design or Design Thinking.

Intrapreneurs are eligible to enter the Business Design stage as soon as the following preconditions are met:

- A clear *value proposition* in a specific field of technology application for a specific customer that has been discovered and partly validated. This means that there is a real customer need and a corresponding benefit can be derived from a Fraunhofer technology.
- A *team* that has shown its willingness and ability to think and act entrepreneurially.
- Considerable *market-based uncertainties* in contrast to pure techno-logical substitutes.
- Technology readiness level: three or higher.

Apparently, there is a seamless transition from Ideation to Business Design. After going through the Business Design phase, the maturity level should look like this:

- The technology exploitation path has to be defined: (1) licensing to incumbent; (2) contract research with industry partners; (3) create a spin-off.
- There is a validated business model tailored to the exploitation path.
- There is a strong team with both business as well as technology expertise that is willing and capable of executing the business model.
- Only a few market uncertainties remain open that can be validated pre-launch (e.g. pricing).

In order to reach that maturity level, Business Design as a process is highly iterative (optimized for fast learning), holistic (we do not leave business blind spots) and focused (uncertainties-driven). Intrapreneurs are guided through the process by three focal elements:

(1) *Four Business Design templates*: first, an advanced business model canvas; secondly, a template to collect analogies, antilogies and major uncertainties as well as to define hypotheses and experiments; thirdly,

a template to plan your very first lean offering; and, fourthly, an action plan to set clear tasks for the team.

(2) *Ecosystem tool*: a tool that assists in sketching the ecosystem of a business. Who are the customers, the end users, the suppliers? Other complementary players or stakeholders? And how are they related?

(3) *Pitch deck*: a professional pitch deck that helps to communicate the status quo to relevant people in a concise and effective way.

Based on these tools, the teams are, on the one hand, able to map the status quo of their business progress at every moment in time without leaving blind spots; and, on the other, they are systematically pushed to learn by focusing on fast validation of critical business uncertainties.

How do we bring this process to life at Fraunhofer? First, we offer a Business Design Game for intrapreneurs and management staff, which is an interactive training format. Secondly, real projects can be pushed forward by our Business Design workshop. Thirdly, and most importantly, we have created so-called 'FDays' ('Fraunhofer Days') – a 12-week internal accelerator program providing intensive coaching to distinguished intrapreneurial teams at Fraunhofer.

FDays have been iterated and adapted several times starting from a mere 'we pay you 12 days for working on your idea with little support during the 12 weeks' (analogous to 3M's and Google's '% of working time' rule) and 'freedom' to a now highly supportive and advanced accelerator program that draws upon world-leading best practices (Techstars, corporate accelerators, Ryerson University, etc.). Iteration by iteration, we have set the bar higher for new entrants to FDays batches (around ten teams) and increased the level of support. Lessons learned from the first iterations were:

- Pure freedom leads to almost zero results, and the researchers remain focused on technological aspects only.
- Too low technology readiness levels cannot nurture the methodological potentials offered by Business Design due to unclear fields of application and a lack of prototype material for customer interaction.
- 'Get out of the building' experiences as well as workshop formats with physical attendance are most effective for transforming researchers' mindsets and pushing the project forward. Bringing Fraunhofer external expertise into the project is crucial.

Due to these lessons learned, the current FDays are structured as follows:

- Kick-off workshop for setting the tone for the 12 weeks, introducing and teaching major tools, motivating the teams and defining the most critical working packages for the first half (six weeks) of the FDays in order to reduce market uncertainty.
- Halfway through the program, FDays Micro Accelerator used to recap the first six weeks, focusing on project work, additional skills workshops and networking with relevant external people. Ends with a major update of the action plan for the second half.
- Demo Day held at the end of the 12 weeks with short pitches on the status quo and major lessons learned. Audience consists of relevant people from Fraunhofer HQ to assure a seamless transition to follow-up support programs.
- Continuous coaching during the 12 weeks by the Fraunhofer intrapreneurship team and its IXperts network (intrapreneur experts) comprising external people with relevant high-tech business expertise such as serial entrepreneurs, investors, innovation managers, etc.

The current version fits seamlessly into Fraunhofer and our researchers' daily environment, provides the right content at the right time and reveals much of the real potential of a project without major financial outlay. Although still learning, we have already found an effective way of translating world-leading best practices in terms of accelerator models, incubators, methodologies such as Lean Start-up or Design Thinking, general innovation management and entrepreneurship in the context of a high-tech research institution.

6.3.2.4 Long-term goals

Intrapreneurial-thinking employees have a major impact on the evolution of an innovative and dynamic organization. They seize opportunities, generate innovations and execute them either internally or externally.

The Venture Lab was launched in early 2013 as a way to stimulate more intra- and entrepreneurial thinking within the Fraunhofer-Gesellschaft. Although an increase in intrapreneurial and entrepreneurial activities has been seen, the timeframe is not sufficient to draw any clear conclusions on a cultural shift. Nonetheless, we are confident of reaching the long-term goals of our intrapreneurship approach as follows.

Fraunhofer intrapreneurs will know and understand the different methods of technology transfer and consider spinning out technologies as a potential career path. Hence, with an increasing number of people aware and willing to create spin-offs, the actual number of Fraunhofer spin-offs will also increase. However, we do not and must not limit ourselves to the individual researcher level. We also need to make sure that the institutes themselves foster intrapreneurship. For us, this means that institutes take responsibility for their own innovation processes and proactively shape and live their own intrapreneurship approaches. We can only 'help institutes to help themselves', meaning that we can only provide support for a certain period of time, until the processes and culture are sustainably changed.

REFERENCES

Blank, Steve (2013), *The Four Steps to the Epiphany* (California: K&S Ranch, Incorporated).

Blank, Steve and Bob Dorf (2012), *The Startup Owner's Manual: The Step-by-step Guide for Building a Great Company* (California: K&S Ranch, Incorporated).

Brown, Tim (2008), 'Design Thinking', *Harvard Business Review*, **86** (6), 84–92.

Keinz, P. and R. Prügl (2010), 'A User Community-Based Approach to Leveraging Technological Competences: An Exploratory Case Study of a Technology Start-Up from MIT', *Creativity and Innovation Management*, **19** (3), 269–89.

Kelley, Tom (2001), *The Art of Innovation. Lessons in Creativity from IDEO, America's Leading Design Firm* (New York: Doubleday).

Martins, E.C. and F. Terblanche (2003), 'Building Organisational Culture that Stimulates Creativity and Innovation', *European Journal of Innovation Management*, **6** (1), 64–74.

Osterwalder A., Yves Pigneur, Gregory Bernarda, Alan Smith and Trish Papadakos (2014), *Value Proposition Designer. Strategyzer* (Hoboken, NJ: John Wiley & Sons).

Pinchot, Gifford and Elizabeth Pinchot (1978), *Intra-Corporate Entrepreneurship, Tarrytown School for Entrepreneurs* (New York: Harper & Row Publishers).

Ries, Eric (2011), *The Lean Startup* (New York: Crown Business).

Wylant, B. (2008), 'Design Thinking and the Experience of Innovation', *Design Issues*, **24** (2), 3–14.

PART 3

Financing spin-offs and technology transfer

7. Incubation to address the 'innovation gap'

Ulrich Mahr and Florian Kirschenhofer

7.1 SETTING THE SCENE

Business incubators support the development of companies, helping them get through the early start-up phase, when they are most vulnerable. They offer their tenants a variety of support services and resources to help them to grow. What started as a stopgap solution in 1959 has turned out to be a great way to meet economic and socioeconomic needs by creating jobs, supporting technology transfer and ultimately driving new industry sectors. It is estimated that there are around 7,000 incubators worldwide. In the US, the number of incubators has increased from 12 in 1980 to 1,250 in 2012 (NBIA, 2014).

Incubators usually apply a screening process to assess the market potential of prospects before allowing them to join the program. Most incubation models focus on start-up projects, while some focus more on technology development. The range of support given to selected and incubated start-up projects is wide and comprises services such as provision of infrastructure, administrative services, network, education, coaching, etc. Business incubation is the first corporate development step in a company's life cycle, as shown in Figure 7.1.

The idea behind incubation is to offer teams the opportunity to concentrate on their main tasks and contribute key resources that a start-up project needs. While many incubator programs do not focus on a specific industry sector, 39% of US incubators target the high-tech field (Knopp, 2007). Time spent in an incubation program depends on the stage of the technology, the type of business and, in the case of a start-up project, the entrepreneur's level of business expertise, and regulations governing the respective incubator. In particular, product and technology oriented companies with significant R&D costs (e.g. in the area of life sciences) require more time and funding in an incubation program than service-based companies with early revenue generation, moderate financing needs and a low-risk profile. The aim of the incubator is to increase the likelihood of a start-up to start

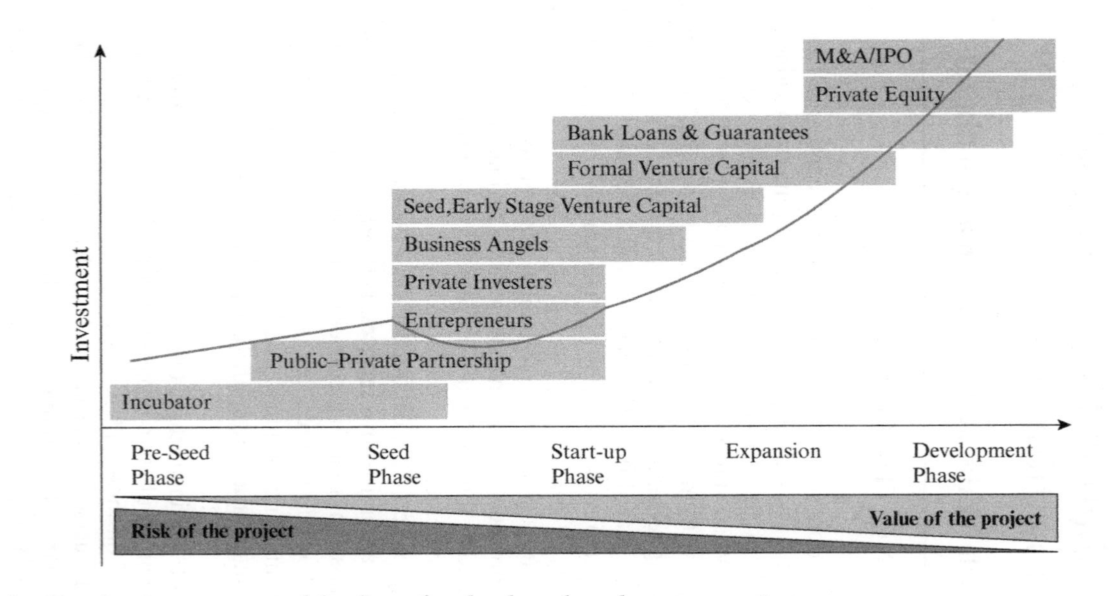

Figure 7.1 Development stages and funding of technology-based start-up projects

and stay in business for the long term after successful graduation from a business incubation program or to commercialize the incubated and validated technology successfully through existing companies.

The first US incubator, probably the first in the world, was established in 1959 in Batavia, New York by chance. A local real estate developer bought a large industrial complex and was unable to find an industrial partner prepared to rent the entire space. He decided to rent smaller units to small firms and provide additional amenities, such as shared office services, assistance in raising capital and business advice to entrepreneurs and small companies (Hackett and Dilts, 2004). The concept worked and others began to copy it. Besides the personal monetary success for incubator owners, researchers and the authorities recognized a positive influence on the economy. More firms with a higher success and survival rate were founded. Subsequently, incubation models were supported and even founded by universities, cities, states and private initiatives across the US.

In the early 1980s, the European Union (EU) acknowledged the need to support entrepreneurial projects and small and medium-sized enterprises (SMEs) by lowering the barriers for entrepreneurs to implement their venture successfully in order to develop EU regions and create economic growth. The EU's solution was to provide knowledge-based services, complementary to the then-existing models of solely supplying affordable office space and services (European Commission, 2010). These complementary services included training and coaching as well as advice to address technical and business development challenges. In addition, these incubators provided access to relevant networks; thus the installation of business innovation centers (BICs) was supported to give entrepreneurs the chance to contact other potential founders, CEOs of small firms, investors, stakeholders and collaborators. Today, there are around 150 BICs across Europe (EBN, 2013). In parallel, European member states, regions, companies, private individuals, universities and research organizations have established additional incubators to push the creation of new businesses.

Israel has also recognized the benefit of providing start-up support via incubators and has successfully incentivized the creation of numerous technology incubators by providing significant public funding. High-tech incubators can be found throughout Israel and are partially funded and administrated by the Office of the Chief Scientist (OCS) of the Ministry of Industry, Trade and Labor. The incubator program was established in 1991 to support the development of innovative technological ideas into viable start-up companies after a two-year period at an 'incubator' (Israel Science and Technology Directory, 2014).

Those incubators have meanwhile spawned a set of highly recognized start-ups and internationally successful market leaders such as Enzymotec

(Naiot Incubator), Compugen (A m-Shav Incubator) and NasVax and Protalix (Meytav Incubator).

These incubation activities have proven to be very successful within an entrepreneurship-friendly environment that provides (i) sufficient funding from venture capital (VC) companies and business angels willing to finance high-tech start-ups from research organizations, and (ii) corporate investors with the necessary knowledge to develop the technologies into products for the market and who are prepared to assume the risk inherent in early stage technology development. Until the 1980s, business incubators were effectively used as a new economic development tool after the stagnation of the US economy in the 1970s. Provision of low-cost services and reduced overhead costs and time devoted to non-value-adding activities were sufficient to stimulate the growth of new companies. Due to increasing unemployment in the 1980s and 1990s in heavy industry sectors such as automobiles, steel production and engineering, the importance of innovation and entrepreneurship for the economy became increasingly clear. Governments committed to supporting entrepreneurship and creating new jobs in future-oriented sectors. Entrepreneurs, venture investors and other private-sector developers started to respond to the expanding small business market. Research facilities at universities and public research institutions started to transfer their research findings and commercialize their intellectual property. This development led to a growth of business incubation programs. Incubation became a popular tool for starting up new companies that lacked management and marketing skills.

During the last decade, the amount of venture capital available for start-up projects in Europe, and especially in Germany, has decreased dramatically. The public market crash in the early 2000s severely impacted the venture capital industry as valuations for technology start-up companies collapsed and the existence of profitable exit opportunities vanished. Ever since the technology slump, investors have drastically reduced their commitments and the VC industry has downsized.

If venture capital investment is measured in relation to GDP, Germany generally ranks on a par with EU countries in financial distress such as Portugal and Greece, and is far behind the northern EU member states of Great Britain and Israel. Especially critical is the situation in the life sciences. Normalized by the respective GDP, the US has approximately eight times and Israel approximately 17 times as much venture capital available compared to Germany (OECD, 2013).

As a result, competition for scarce financial resources has become fierce and requirements for financing have risen substantially. In order to prepare for successful fundraising, start-ups not only need to assemble strong and experienced management teams, but also provide sufficient supporting

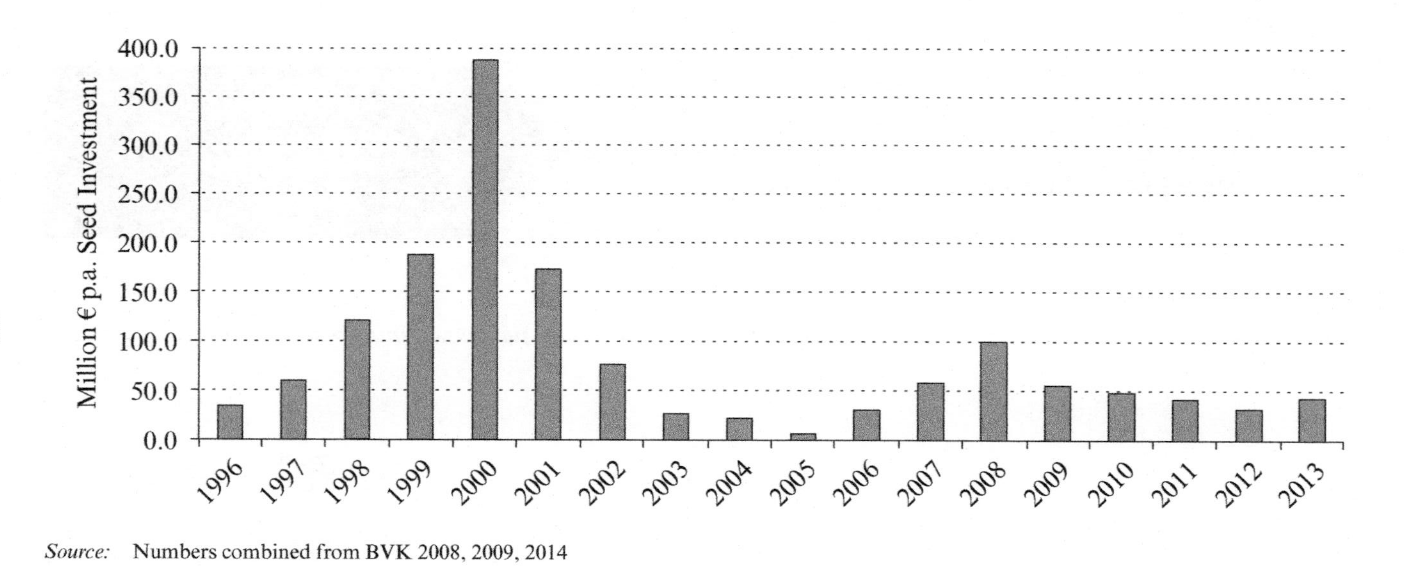

Source: Numbers combined from BVK 2008, 2009, 2014

Figure 7.2 Venture capital investment in start-ups in Germany

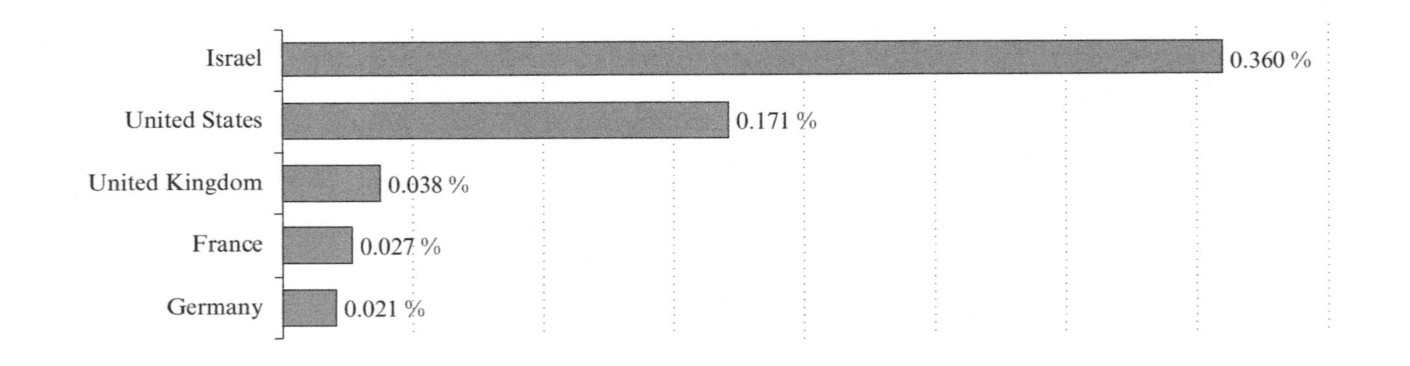

Source: OECD, 2013

Figure 7.3 Total venture capital investment (seed-, early- and late-stage) in relation to the GDP of the specific country in 2012

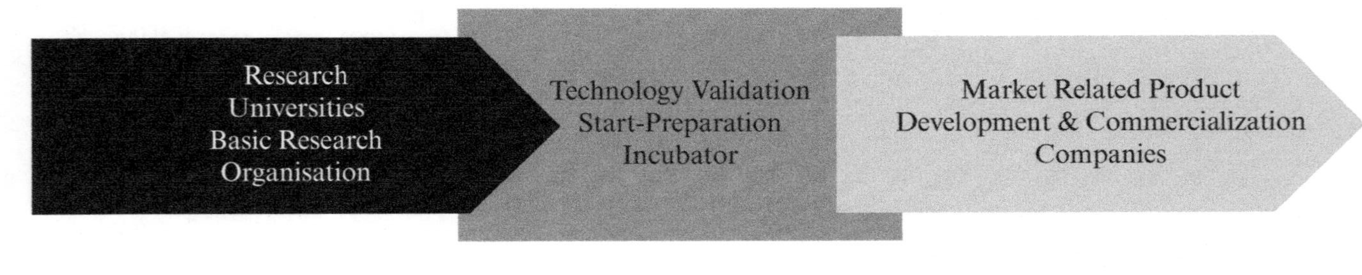

Figure 7.4 The incubation process

data in an industry-compatible manner for their intended innovation. This is typically difficult to achieve in a pure research environment. The primary aim of research organizations and academic scientists is to generate fundamentally new insights, which are presented at renowned scientific conferences and/or published in high-ranking scientific journals. This mission is often fulfilled through proof of principle experiments in scientific model systems. At this stage, the technology is usually only demonstrated within the research group based on a single experiment and not necessarily in an industry acknowledged 'gold standard' model system. However, numerous inventions that are principally suitable for commercial exploitation are made at universities and research institutions every year in connection with basic research.

Many established companies are also no longer willing to in-license early stage technology projects from research organizations or technology-based start-ups without thorough characterization and validation leading to a similar gap. In addition, some of them no longer have the resources and in-house capabilities to develop early stage technologies into marketable products.

The lack of funding and industry appetite for early stage technologies leads to the so-called 'innovation gap' between the scientific findings and their technological application, in particular for those technologies that require further market-oriented development.

Without appropriate measures to tackle this problem, the innovation gap will lead to a situation whereby promising research results with relevant application potential, which were previously financed by the taxpayer, cannot be further developed and offered as products or services on the market.

Existing incubation approaches of technology and start-up centers, like BICs, that only provide suitable and affordable infrastructure and advice cannot address this innovation gap.

Public authorities, and technology transfer organizations like Max-Planck-Innovation GmbH, have been developing and introducing new concepts and initiatives to overcome this. These aim to address the requirements of investors and licensees alike by (i) reducing the inherent risk residing in early stage technologies; (ii) providing infrastructure and a mixture of private and public funding for the incubation phase as well as the later start-up phase; and (iii) providing coaching, administrative and planning support during incubation and preparing the actual spin-out or out-licensing process.

Risk reduction is achieved by generating additional confirmatory proof of concept data according to industry standards. The relevant set of R&D experiments are defined prior to the actual start of incubation with input from industry experts and key opinion leaders. The R&D plan is later executed and eventually adapted as needed during incubation.

Funding for the incubation typically comes from both public and private sources coordinated by the incubator. The R&D and office infrastructure is established and provided by the incubator with public support in a suitable environment, such as established research centers, in the respective field.

If the technology is commercialized through a start-up, the founding team will be coached to develop the skills necessary to implement the new company successfully. In most cases, the team is supplemented with new, experienced members with complementary skill sets or the incubator establishes a completely new team. Identifying suitable and qualified persons to add to the management team and coaching the existing team are, in addition to the technology development, crucial factors for a successful incubation of technology-based projects from research organizations. Furthermore, the incubator should support industry-compatible R&D planning, provision of relevant expert input, administrative support, business planning, business development support and, last but not least, fundraising support and eventually co-financing resources to facilitate the syndication of investors.

After the incubation phase, the projects should reach a development stage that matches the demands of third-party investors or industrial partners.

Three examples of such new incubation concepts – the Life Science Inkubator in Bonn and Dresden, the Photonik Inkubator in Göttingen and the IT Inkubator in Saarbrücken – demonstrate the relevant components needed to efficiently address the 'innovation gap'.

7.2 LIFE SCIENCE INKUBATOR (LSI)

Based on a unique and comprehensive approach in Germany, LSI incubates selected, high potential start-up projects in the life sciences (biotech and medtech). Project proposals undergo a rigorous due diligence assessment of the technology, intellectual property, market, competition and the team itself. Selected and incubated projects receive appropriate funding for the validation of the respective technologies as well as a fully equipped lab and office space. In addition, the teams are coached and complemented by industry experts, LSI's project management and experienced interim managers with regards to both planning and execution. LSI also supports business planning and fundraising. If a suitable lead investor is identified, LSI can provide additional co-investment from its associated fund vehicle (LSI PSF GmbH).

LSI carefully selects incubator candidates in a multi-step due diligence process. First, project proposals are identified that generally match the requirements and criteria established by LSI. These projects then undergo an initial internal assessment by LSI project management, including meetings with the teams. If the outcome is positive, a challenge workshop with experts in the field is organized to address critical issues and challenges of the project and the team. Furthermore, an external expert assessment provides guidance to the management and investment board regarding the final decision as to whether to incubate the project. The external assessment comprises an evaluation of the science, market and intellectual property, and includes a team assessment. The proprietary assessment center was developed by LSI together with a partner recruitment firm and aims to identify teams with coaching potential as well as gaps that need to be filled by external management resources.

LSI and the respective project team agree on a suitable R&D plan with relevant milestones to be achieved during incubation. The implementation of the R&D plan is monitored concurrently and necessary adaptations to achieve an attractive industry compatible data set will be incorporated from time to time as needed. If external input is required, LSI will help to identify and hire an experienced expert.

To prepare the actual spin-off and fundraising, LSI supports the development of a professional business plan. For that purpose, LSI can hire an interim manager to coordinate the respective activities and start business development and fundraising activities. LSI has an established network of renowned venture capital companies to support fundraising and LSI involves the venture capital firms throughout the incubation phase and eventually also in the selection phase to maximize the chances of subsequent financing.

In April 2014, the first spin-off 'Neuway Pharma GmbH' received its initial (Series A) financing from the renowned lead investor Wellington, kfw, private investors and LSI's co-investment vehicle LSI PSF. Another spin-off, 'Bomedus GmbH', was able to attract venture funding of approximately 2.5 million euros following its earlier seed financing.

LSI is structured as a private–public partnership with funding from public sources (federal Ministry of Research and Education (BMBF), regional ministries of North Rhine Westphalia (MIWF) and Saxony (StMWK)). Apart from the hosting Center of Advanced European Studies and Research (Caesar) in Bonn, the renowned Max-Planck-Society, the Fraunhofer Society and Helmholtz Association became partners and shareholders of LSI. LSI PSF provides co-investment funding with its major shareholder NRW Bank and contributions from other regional banks or investors as well as the research institutions. LSI is comprised of

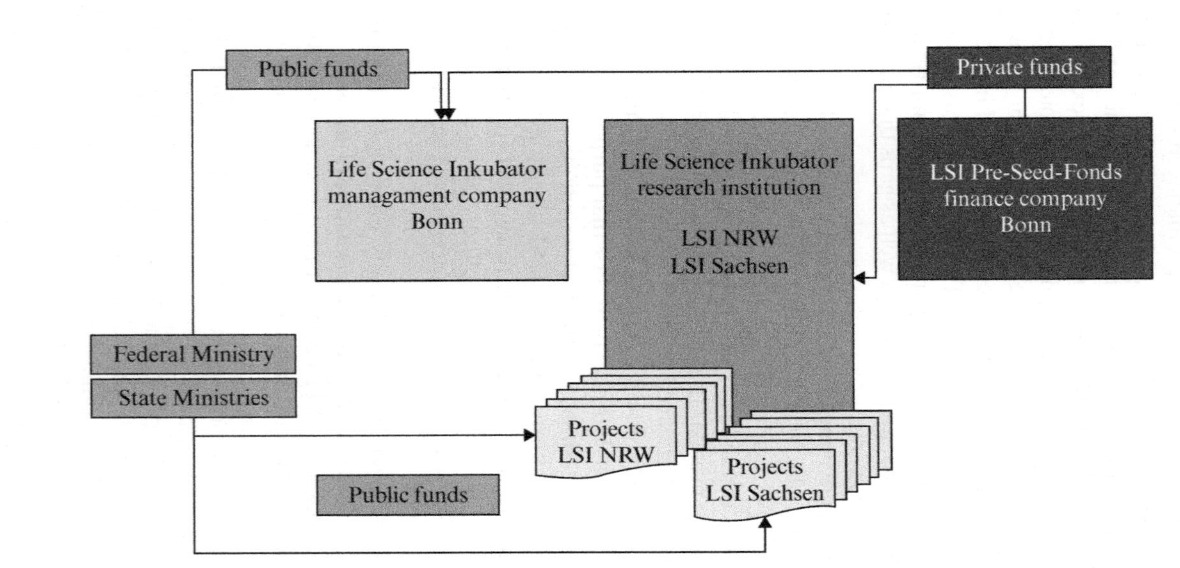

Figure 7.5 Structure of LSI: the public–private partnership combines management, research and finance

a management company (LSI GmbH), the actual incubator (LSI KG) and the co-financing vehicle (LSI PSF GmbH).

In return for its financial support, the LSI PSF has the right to establish and participate in the spin-offs incubated by LSI with a pre-defined fixed participation scheme.

Based on the successful establishment of LSI KG in Bonn, LSI Sachsen KG (a regional subsidiary of LSI) was established in January 2013 in Dresden with strong support from the Saxonian Ministries of Education and Science as well as the Ministry of Economy.

7.3 PHOTONIK INKUBATOR (PI)

As a consequence of the successful implementation of LSI and a positive evaluation by a renowned international advisory firm, the applicability of this incubator model in other industries with key enabling technologies was assessed. LSI's innovative approach to fostering innovation via new start-ups encompasses several tools that can be applied independent of a specific industry. Therefore, to prove the functionality of the incubation approach for further technological fields, an ideal industry had to be identified. Given its explicit listing in the catalogue of key enabling technologies of Germany's high-tech strategy, and both the technological and market strength of photonics in Germany, photonics was chosen. Photonics is one of the strongest economic industries in Germany and yet the industrial success of optical technology plays only a limited role in the spin-off activity from universities and other research institutes. Due to its outstanding optical infrastructure and photonic strength, the Laser-Laboratorium Göttingen (LLG) in lower Saxony was selected as the location for PI. The technical expertise at LLG and its established network of industrial partners provides an excellent basis for cooperation and support.

At LSI, the management has put considerable effort into identifying management candidates to complement the experience of founding teams in areas where they lack expertise, such as project management and general management. Based on such a review, if necessary, decisions are made to assemble an entirely new executive team during incubation and after spinning-out the projects ('Gründen ohne Gründer'). The intention is to allow PI to actively focus on identifying suitable technologies with an attractive profile, rather than being tied to a specific founding team, and the possibility to add an experienced management team to advance a promising technology into a start-up that is professionally prepared and managed. During evaluation, the identified projects are conceptually (re-) designed with respect to the market, technology, IP and business model.

While incubation projects at LSI are required to move into the respective site and center their development efforts at the facilities of LSI, incubation projects at PI enjoy greater flexibility with regard to the execution of R&D projects, especially if they require substantial equipment expenditures ('decentralized incubation'). A novel module will also be implemented to address the possibility of multiple utilizations of technologies and IP. The complete range of best practices for successful incubation will be applied at each technological and regional incubator.

To date, PI has reviewed more than 70 different project proposals; the first incubation project began in November 2014. The intention is to secure private funding for the establishment of a pre-seed fund, which will provide co-financing resources for general PI operations and also co-investment in spin-offs comparable to the LSI PSF model.

7.4 IT INKUBATOR SAARBRÜCKEN (ITI)

The focus of ITI is on the incubation of projects in the field of information technologies. It was founded in 2013 in Saarbrücken as a subsidiary of Max-Planck-Innovation GmbH and the knowledge and technology transfer organization of the Universität des Saarlandes Wissens- und Technologietransfer GmbH. Software projects from all Max-Planck institutes, the Saarland University and Saarland universities of applied science can be considered for an incubation phase. Because there is already efficient coverage of the internet space by private incubators, business angels and venture capital firms, there is no need to provide additional support for these projects. They are therefore excluded from the scope of ITI.

Teams applying for an incubation phase generally work in a special scientific environment. In this environment, inventors and scientists are able to work with the software code they have developed, but when entering ITI, the software is generally still too complicated and underdeveloped for a broader, consumer-based use of the technology. Industry-related interfaces, user-friendly graphical user interfaces, code documentation and optimization of the program to increase speed and decrease size are usually missing. Furthermore, the use of open-source components is prevalent and could be an obstacle for commercialization. Following incubation, projects should be developed to an industrial level from which they can be commercialized, either by licensing to existing firms or as start-ups.

In principle, three different types of projects can be incubated:

(1) *Start-up projects*: Teams working on promising IT projects can apply for start-up incubation if they plan to establish a company after the

incubation phase. The technological status and the team are evaluated by ITI. Assistance and coaching for the team is provided and paid for by the incubator. If additional expertise is needed to build a promising team, appropriate individuals can be identified and hired. Office space, technical infrastructure and access to outside technical experts are provided and financed by ITI. In addition, the team is coached and supported to develop a business strategy and plan. The team should develop strong and weaker ties with a network of potential partners during the incubation period with support from the incubator in order to overcome potential weaknesses due to the firm's small size and early development stage (Aldrich and Auster, 1986). Network contacts should be developed before they are needed, thus network building is essential for entrepreneurial teams.

If a new company is established, it needs a license to use and commercialize the technology. Usually the start-up is founded by the incubation team members at the end of the incubation period and is financed by revenues, the founders, or external investors in combination with public support programs. The license contract must include 'market-based terms' to ensure that small firms do not receive preferred conditions over larger, existing competitors. Therefore, the contract must contain different payment streams, such as a down payment, milestone payments, and royalties based on the revenues generated with the licensed technology. However, most small firms suffer from a lack of resources, especially financial resources (Cooper et al., 1991). Investors want to finance the further development of new technologies, but not necessarily down payments for licenses from public research organizations. The solution is to convert early license payments, such as down payments, into shares in the new company. This results in the incubator as well as the research organization becoming shareholders of the firm.

(2) *Technology development projects*: Some promising technology projects from research organizations are either unsuitable as start-ups or no team can be formed to realize a new venture. A project is unsuitable as a start-up if the market structure and market powers are not supportive of small firms. In such an environment, the start-up would be unable to generate enough revenue to become a profitable venture or to find an investor if additional funds were needed to implement the entrepreneurial venture.

In these cases, ITI project managers assume responsibility for incubating the technologies. They can cooperate with external freelancers or companies and the researchers who originally invented the technology at the research organization. At the end of the incubation phase, the responsible incubation managers, with support from the ITI general management and the technology transfer organization of the initial research organization, start searching for suitable industrial partners for the further development of the project or to commercialize the technology directly to customers and/or consumers. In most cases, the technology is licensed to an existing company, which pays customary license fees for access to the technology.

(3) *Start-up without a team*: In cases where technologies meet the criteria for the establishment of a start-up, but none of the inventors are interested in developing a new venture, the incubator can organize the further development of the technology and build an entrepreneurial team. The team should establish a new venture after the incubation period and, from that point, it is treated as a start-up project. The incubator receives a significantly higher percentage of shares in the start-up than in other ventures because it is fully coordinating the initiation and eventual success of the new company.

The typical incubation phase of an IT project at ITI lasts from six to nine months. This is significantly shorter than in other industries, such as life sciences or photonics, and also less resource-consuming.

ITI employs only a small team of incubation managers and team assistants. The in-house experts are responsible for the analysis of the technical needs and status quo of the project and to complement the experience of team members. They develop an action plan in collaboration with the CEO and project team to overcome any technological gaps and to support team development. Technical problems are solved by external experts and/or project teams, who are usually the researchers who invented and developed the technology. ITI management and external coaches support the development of the internal team.

ITI's lean structure combined with the use of external experts enables the incubator to develop very heterogeneous technologies. If necessary, special tasks are fulfilled by external partners and subsequently integrated into the project. The core know-how is concentrated at and controlled by the incubation team. According to German law, the employer at the time of invention owns the technology. Therefore, if new technologies are invented during the incubation phase, the incubator will own the intellectual property rights. If a start-up is created, its team has the right of first refusal to in-license the technology at market rates after the establishment of the new company.

The decision to commercialize the technology via a new business or to license the technology to an existing firm is deferred to a later stage in the development process. The team evaluates both options at each stage of incubation in order to find the best solution for every project. Even though the average incubation period of six to nine months is relatively short, this mechanism is necessary to keep pace with the fast-changing IT environment.

The government of Saarland financed the initial start and growth phase of ITI. However, venture arms of local banks are showing increasing interest in creating a public/private partnership to finance additional incubation projects and start-ups based on ITI technologies.

7.5 KEY SUCCESS FACTORS

Methods and approaches had to be developed to overcome the increasing 'innovation gap' and, combined with existing support, potentially to facilitate technology transfer and foster innovation via licensing to existing companies and especially to start-ups.

The following crucial components of successful pre-seed incubation have been identified:

(1) *Deal flow*: First, an incubator needs to develop an adequate deal flow of sufficiently high-quality projects. This requires significant scouting activities in the different target groups and an intensive public relations campaign to promote the offer of the incubator. Frequent interactions with potential investors and tech-transfer organizations can also yield good sources of high quality but early stage opportunities requiring incubation.

(2) *Assessment/selection process*: The pipeline of incubation candidates needs to undergo a rigorous due diligence process to ensure the quality and potential of selected candidates. Due diligence typically focuses on team assessment, technology and innovation, intellectual property, and market and competition. Internal assessment by the incubation management team should ideally be complemented with outside experts. The final decision regarding the actual selection of start-up projects should be taken by an independent expert jury based on (i) an investment proposal by the management and (ii) a team presentation.

(3) *Team audit, management and complementary experiences*: The consensus amongst investors in start-ups such as venture capital companies is that the most prominent key success factor in a start-up is an experienced management team with complementary skills and know-how. Only in rare instances do start-up projects from universities and research organizations have such a team in place prior

to financing. In most cases, teams lack industrial experience and know-how in industry-like R&D planning, project management and general management. Therefore, coaching, external consulting by industry experts, team building through the integration of additional team members and involvement of external, experienced (interim) management are very important factors in preparing a successful spin-off. A professional recruiting process in close cooperation with existing team members is a prerequisite for building a successful and strong heterogeneous team.

(4) *Pre-seed funding*: Funding at universities and research organizations is often scarce and restricted to basic research. But sufficient funding is needed to address the requirements of industry and investors alike regarding the validation of the technology in question. Ideally, all the financial requirements for preparing a start-up should be addressed by an incubator in a highly flexible manner. This means not only covering the cost of suitable offices and labs but also funding necessary R&D work to validate the technology, including staff costs, capital expenditures, consumables, travel expenses, etc.

(5) *Project management*: The way R&D is conducted in the academic environment is different to industry. While new discoveries motivate swift changes in research directions in academia in order to elucidate new findings, industry's primary interest is to define an R&D goal and attempt to achieve that status as soon as possible within a given budget and regardless of what interesting findings may occur during that phase. The switch from one approach to another requires time and constant support. Project managers need to check-in at short, regular intervals to determine whether the predefined R&D development goal is on track and adjust the planning if necessary to achieve it.

(6) *Infrastructure*: High-tech projects from research organizations are complex and usually highly specialized. Access to internal and external infrastructure is typically needed to further develop the technology at industrial level. Due to the heterogeneity of projects within certain fields, there is often a need for diverse and expensive specialized equipment, most of which the incubator may not have available in its own infrastructure. Cooperation, contracting and renting are possible ways to ensure access to these essential resources.

(7) *Administrative support*: Ideally, founding teams should focus on advancing the technology through validation and data generation as well as business planning, rather than administrative tasks. If incubators take over administrative tasks, innovation can advance much faster and with less friction.

(8) *Networks*: Even the best project cannot survive or grow in isolation. Both strong and weak ties with a network of different stakeholders, which may change over time, are essential for the success of entrepreneurs and their start-ups (Witt, 2004). Incubators help to build and grow these networks of potential partners, investors, service providers, etc. Most new entrepreneurs have never started a venture before; therefore, establishing contacts with other entrepreneurs, who are in the same situation, can be beneficial.

(9) *Financing*: The funding situation for start-ups is critical in Europe and especially in Germany. The number of venture capital companies is decreasing and banks rarely provide financing to start-ups. Co-financing by the incubator can significantly support fundraising efforts because it signals confidence in the start-up and reduces the total amount of additional external financing needed to start the business.

(10) *Post incorporation support*: While securing financing is extremely challenging, starting a business without experiencing significant delays is almost as challenging. To ensure a smooth transition from incubation to start-up, the option of rental agreements covering lab space and instruments can be valuable. The availability of a well-established network of service providers for outsourceable office functions, such as bookkeeping and auditing firms, lawyers, patent attorneys, marketing/PR, telecommunications, etc., is also a clear accelerator. Cooperation with existing accelerator programs (e.g. German Accelerator) can be a very useful addition.

In conclusion, the following success factors should also be addressed when developing the overall concept of incubators and incubation.

An appropriate infrastructure and sufficient funding for proof of concept R&D work and narrow cooperation and interaction with potential investors to minimize the time gap between the end of incubation and follow-up funding are important. The set-up of an incubator should facilitate the transition from publication-oriented research to industry-oriented R&D during incubation and before the actual spin-off is established. This is best achieved by transferring the team from its academic environment to an outside incubator combined with intensive advance planning and selection process coordinated by the incubator's project management supported by external industry experts. Knowledge of industry requirements and continuous monitoring and support during the incubation process are a prerequisite of the successful preparation for a later spin-out. Last but not least, the experience and commitment of incubator employees is essential. Ultimately, it is they who will guide the teams and projects from science to success.

REFERENCES

Aldrich, H. and E. Auster (1986), 'Even Dwarfs Started Small: Liabilities of Age and Size and Their Strategic Implications', *Research in Organizational Behavior*, **8**, 165–98.

BVK (2008), 'BVK-Statistik – Das Jahr 2007 in Zahlen', http://www.bvkap.de/media/file/163.BVK_Jahresstatistik_2007_final_210208.pdf (accessed 15 January 2016).

BVK (2009), 'BVK-Statistik – Das Jahr 2008 in Zahlen', http://www.bvkap.de/media/file/222.20090309_BVK_Jahresstatistik_2008_final_Lang.pdf (accessed 15 January 2016).

BVK (2014), 'BVK-Statistik – Das Jahr 2013 in Zahlen', http://www.bvkap.de/media/file/501.20140224_BVK-Statistik_Das_Jahr_in_Zahlen2013_final.pdf (accessed 15 January 2016).

Cooper, A., J. Gimeno-Gascon and C. Woo (1991), 'A Resource-Based Prediction of New Venture Survival and Growth', *Academy of Management*, Best Paper Proceedings, 68–72.

EBN (2013), *EC-BIC Observatory 2013 and the Last Three-year Trends* (Brussels: European BIC Network).

European Commission (2010), *The Smart Guide to Innovation-Based Incubators (IBI)* (European Commission: Publications Office of the European Union).

Hackett, S. and D. Dilts (2004), 'A Systematic Review of Business Incubation Research', *The Journal of Technology Transfer*, **29**, 55–82.

Israel Science and Technology Directory (2014), 'Technology Incubators', http://www.science.co.il/Technology-Incubators.asp (accessed 15 January 2016).

Knopp, L. (2007), *2006 State of the Business Incubation Industry* (Athens: NBIA Publications).

NBIA (2014), 'National Business Incubation Association', http://www.nbia.org/resource_library/faq/#3 (accessed 15 January 2016).

OECD (2013), *Entrepreneurship at a Glance 2013* (OECD Publishing).

Office of the Chief Scientist, Ministry of Economy, Tel Aviv, Israel (2014), Technological Incubators Program Office 2014, http://www.incubators.org.il/category.aspx?id=606 (accessed 15 January 2016).

Witt, P. (2004), 'Entrepreneurs' Networks and the Success of Start-ups', *Entrepreneurship & Regional Development*, **16**, 391–412.

8. The seed challenge

Michael Brandkamp

8.1 INTRODUCTION

In 2014, the Ice Bucket Challenge excited the world! According to Wikipedia, it 'encourages nominated participants to be filmed having a bucket of ice water poured on their heads and then nominating others to do the same. A common stipulation is that the nominated participants have 24 hours to comply or forfeit by way of a charitable financial donation' (Wikipedia, 2016a).

It was a very popular, refreshing and exciting experience. The goal was to raise funds for research into the disease ALS (amyotrophic lateral sclerosis). Although there are quite a number of differences, investing in new technology-based firms (NTBF) is also very refreshing and exciting as well as challenging.

The German born Peter Thiel, who is worth US$2.2 billion, ranks number four on the 2014 Forbes List. He was the first outside investor to buy 10.2% of Facebook's company shares for US$500,000. Prior to that, he co-founded PayPal. This company was sold to eBay for US$1.5 billion. At that time, his remaining 3.7% stake was worth approximately US$55 million (Wikipedia, 2016b).

This is not really a German story, since Peter Thiel moved to the US when he was just one year old. However, Germany also has some great examples of investors taking high risks and ending up with great returns. Falk Strascheg, who founded, developed and sold a company dealing with laser technologies, invested his money in young, high-potential, high-tech companies. He founded the most successful German venture capital (VC) company called Technologieholding, investing very profitably in early stage companies like Intershop, Mobilcom and EOS. Many others were sold successfully on the 'Neuer Markt'. Technologieholding itself was acquired by 3i Group in 2000 for a total of US$155 million (Bloomberg, 2015). This has made Strascheg a very rich man.

Today the start-up scene is booming in Berlin, where the German start-up association estimated that there were 2,500 young companies in 2013

(Benrath, 2013). The Berlin start-up scene attracts young people with great potential from all over the world, who join the German ecosystem as entrepreneurs or managers. There are also a considerable number of incubators, accelerators and business angels offering both money and valuable business expertise and support (Hansen, 2014).

Berlin has already produced several start-up success stories, of which prominent examples include Rocket Internet, Zalando, Lieferheld and Mister Spex. Of course, this number remains limited. However, many of the young companies in the city follow Eric Ries' Lean Start-up concept (Ries, 2008). They try to build up their companies with very small external investments contributed by family and friends, generating income through selling products and services. In addition, they learn what their customers really need and what they are ready to spend their money on. The initial investment needed for these sorts of lean start-ups is relatively small, especially in the internet business. To set up an internet shop, the initial investments are minimal. Capital is needed for marketing; for example, search engine optimization (SEO), a fulfillment center and a drop shipper to deliver the goods to the consumer.

A number of experts and politicians are now changing their investment focus from the seed market towards the later stages of the value chain for building up NTBFs. So, is the seed market still an attractive area in which to invest?

8.2 SEED CAPITAL MARKET AT A GLANCE

Traditionally, the seed market is regarded as a very difficult market in which to do business. The numbers (Figure 8.1) clearly demonstrate the turbulent nature of investment cycles. After a boom in seed investments in 2000, the market dried up almost completely. At this peak of activities you could also find a large number of incubators and accelerators. Many of them were backed by large corporates. Most of them disappeared after 2000 because of very poor returns. Quite a number of early stage VC companies left the space due to high failure rates.

The market recovered after 2005 with the help of funds backed by the public sector, ready to invest at the seed stage. Globally, the seed capital market is supported by government-related activities. Even the US has a support scheme in place called SBIC, a multi-billion dollar program founded in 1958 (Small Business Administration (SBA), 2016).

It is a similar story in Israel – the so-called 'Start-up Nation' – with YOZMA. In 1991, the Israeli government created 24 technology

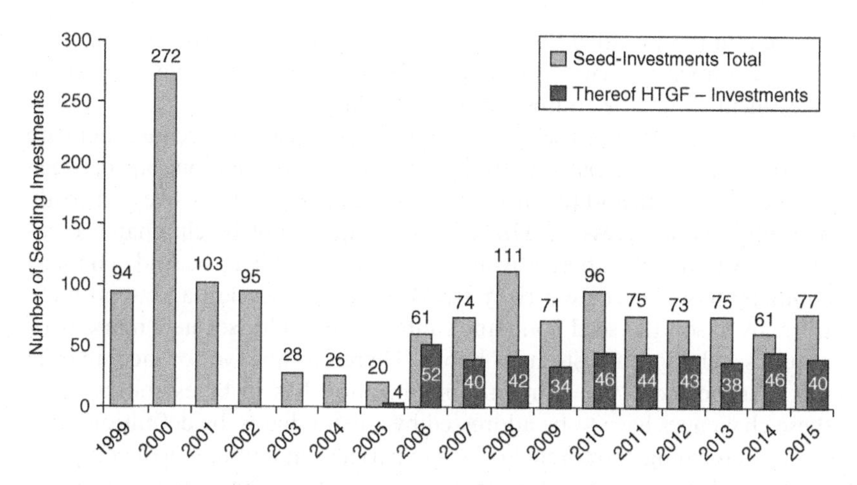

Figure 8.1 Seed capital investment cycles in Germany

incubators. At the same time, the Ministry of Finance founded YOZMA, which means 'initiative'. YOZMA was a fund-of-fund and co-investment scheme with finance of around US$100 million for jumpstarting the development of both the VC industry and start-ups in Israel. In *Start-up Nation: The Story of Israel's Economic Miracle*, Dan Senor and Saul Singer quoted Orna Berry, an Israeli scientist, high-tech entrepreneur, businesswoman and chief scientist and head of the industrial R&D operation of the Israeli Ministry of Industry, Trade and Labor (1996–2000): 'On the success of venture capital and high-tech entrepreneurship in Israel, to paraphrase Lennon (who said "Before Elvis, there was nothing"), before YOSMA, there was nothing' (Senor and Singer, 2011).

However, considering the big success stories of Peter Thiel, Falk Strascheg and the Samwer Brothers, why is there a need for government intervention in the first place?

8.3 DIAGNOSIS OF MARKET FAILURE

It goes without saying that a polar bear cub cannot survive in the North Pole on its own without the protection of its mother. In such an inhospitable environment, the cub needs warmth and nurturing. Similarly, a nursery for trees provides seedlings with a specially protected environment in which to grow. There is similar understanding among policymakers and experts looking at seed-stage companies. Bonaccorsi and Montaina (2012) stressed, in the so-called Policy Brief No. 5 of the European Commission, that 'Venture capital should be mainly left to the market, while seed capital investment should be addressed by the public sector'.

The reason is that investing at seed stage is challenging because there are several considerable risks: there is typically an information gap between the founding team and the investor. To reduce this, an intensive and costly due diligence is necessary. However, the gap cannot be eliminated completely. Additionally, there is usually no product ready to be sold. The seed company has to execute a risky R&D project to develop a new product. In many cases, the seed company's innovation addresses new markets or requires a change in customers' habits. Therefore, an investor faces market risk. Furthermore, the company organization has to be established. All these challenges have to be addressed by a team that is, by definition, new. Very often young entrepreneurs lack experience in building up a company. Even serial entrepreneurs find themselves in new teams. The performance of these teams is hard to predict because a company history is not available. Alongside management uncertainty there is a financial issue. Because most NTBFs need much more funding than a seed investor is able to commit, follow-on investors are needed. In some cases they lack regard for the seed investor's intentions.

Banks are obviously not the right institutions to carry these risks, particularly because NTBFs cannot offer any collateral.

Even VCs cannot necessarily calculate the risks of a company at seed stage. Therefore, most private investors stopped investing in very early stage companies after the dot-com bubble burst in 2001. They moved out of the seed market because of both high and barely calculable uncertainties and bad returns.

Companies able to acquire sufficient capital are an important exception. Only a few have the chance to take off. However, taxpayers need more NTBFs because they invest billions of euros in science, especially in Germany. There are institutions like Max-Planck, Helmholtz and Fraunhofer alongside a large number of universities. Disruptive new ideas are rarely commercialized, although they are very important for the global competitiveness of an economy. The German economy, having very few raw material resources, relies particularly heavily on innovation.

A new company faces many obstacles at the seed stage: lack of capital and a lack of entrepreneurial spirit and skills are regarded as the most prominent. Because of that, the German government and some federal states have implemented seed funds. The purpose of these seed funds is to bridge the seed capital gap in order to establish so-called venture capital readiness in start-ups.

What is needed to develop a high-performing seed capital market to build up a sufficiently high number of NTBFs? To answer that, we need to understand the factors and boundary conditions responsible for the success of a high-tech start-up.

8.4 SUCCESS FACTORS FOR NTBFS

If asked about the crucial issues he/she looks for prior to investment, an investor will probably point out three pillars of success:

(1) *The business idea*: This includes the underlying technology (well-protected by patents or other means) to develop innovative products or services. There must be a clear unique selling proposition (USP) that is difficult to copy and offers a big advantage over existing and future competitors. The NTBF should address niche markets with high growth potential, so that it has a chance to grow with the market instead of fighting the competition. Investors are also interested in very new approaches to address customer needs. In a nutshell, they are looking for a game-changer.

(2) *The team in charge of executing the business idea*: Professional investors in NTBFs agree that the team is most responsible for the success or failure of the company. Therefore, the best entrepreneurs and managers are needed to develop a young firm. Since building up a high-tech company is a very complex and challenging task, a team of more than one but less than four members is preferred, comprising both high-profile technology and entrepreneurial leaders. The team is not only the most crucial pillar for success; it is also the most difficult one to evaluate. Most investors lack a reliable tool to assess teams. They rely on individual judgment instead of professional tools and prefer working with known serial entrepreneurs.

(3) *The business concept*: This shows how the team would like to execute the business idea. Investors know that the business plan does not reveal the precise picture of the NTBF's future development. However, it does contain the strategy and goals on which the shareholders and managers of a start-up agree. It demonstrates, on the one hand, the opportunities and relevant profit and growth perspectives and, on the other, the risks, challenges and obstacles the company will probably face. It explains how the managers want to organize upcoming tasks, e.g. team, supply management, finance, controlling and sales. The experienced investor is aware that there will inevitably

> be many changes along the way. Therefore, all shareholders and managers have to be flexible. One main advantage of start-ups is the versatility and flexibility to learn and react fast.

However, honest seed investors admit that they do not know everything about success factors. They are not always able to separate successful NTBFs from failures. Many different empirical studies provide data on the correlation between success and factors like age, gender, founder's experience, size of team or financial resources (Boyer and Blazy, 2013). At least 30% of new companies – in many cases far more than that – will fail despite a seed investor's extensive due diligence.

All things considered, it remains extremely difficult to predict the success of an NTBF because there are so many external factors including framework conditions and luck. Luck does not equate to destiny. Luck is determined through level of activity. The more proactive a person is, the better his/her chances of making the right connections. These connections have the power to change the entire story.

However, luck is not determined just by the activity level and quality of the management team. The probability of finding a high-level connection makes a real difference. It depends on both the quantity and quality of the individuals and institutions a management team will meet. This defines the power of the business ecosystem for NTBFs. It should offer a hospitable environment in the same way as the mother polar bear offers warmth and nurturing for her cub in a region of permanent ice.

8.5 SUCCESS FACTORS FOR BUILDING UP A WELL-PERFORMING SEED CAPITAL MARKET

If founders are asked what they need, they will probably talk about capital. Indeed, liquidity is crucial because companies go out of business if they run out of cash. To found a growth-oriented company, it is usually necessary to obtain external funding. In accordance with European Commission guidelines, member state governments have set up seed funds. The largest and most active of these in Germany is the High-Tech Gründerfonds (HTGF). HTGF is publicly backed by the Ministry of Economic Affairs and Energy and KfW, a government-owned development bank, and privately backed by large corporations that want to work with NTBFs. In addition, the federal states have established a number of seed funds, of which most collaborate with HTGF. Alongside these funds, there are an increasing number of business angels, incubators and accelerators across Germany.

However, capital is just like food for the polar bear cub. It is important, but by no means sufficient. The cub needs much more than this to survive.

At the HTGF conference in 2012, James Wong, partner of Garage Venture, responded to the question 'What makes the start-up system in the Bay Area around San Francisco so successful?' with this answer: 'Events like the Family Day 2012, and the ecosystem that comes with it makes all the difference' (Wong, 2012). HTGF agrees. It delivers 'added value' to improve the ecosystem for start-ups.

So what is a good role for an investor in building up an NTBF? Is it his/her responsibility to work on the ecosystem for start-ups? Not really. But a good investor is active and supportive. He/she shares the same entrepreneurial spirit and is ready to support the company as much as possible. The ecosystem is the result of the investor's added value and entrepreneurial activities in and around start-ups. To summarize: good VCs contribute significantly to the establishment of an ecosystem.

Christoph Janz, managing partner at the venture capital firm Point Nine Capital and previously a successful serial internet entrepreneur and angel investor, presented in his blog a very convincing and clear analysis of how to distinguish between a good and a bad investor (Janz, 2014). This blog is highly recommended both for entrepreneurs to judge the quality of investors and for investors to benchmark their own activities.

The good investor plays an active role in supporting the start-up team wherever he/she can but without interfering in the execution of company operations. He/she acknowledges the company management as specialists in specific issues relating to operations of the portfolio company. However, the good investor transfers to the company the experience he/she has gained from building up his/her own companies, contributes his/her network to connect the managers with relevant experts, advises the company, helps to find additional managers if needed, and connects the company with new investors and big corporations. He/she knows about the challenges that must be overcome in order to make an NTBF a large company and works with the NTBF's team to find the best solutions instead of presenting his/her idea as the only sensible one.

Building up a new high-tech firm is a very complex process, which up until now has not been fully understood. The NTBF operates in a very complex system. However, as long as we continue to create highly relevant networking possibilities for the NTBF, promising structures in those very complex systems will probably appear. This means that we do not have to understand the single factor responsible for success. If we create an ecosystem offering high potential and highly relevant connections, companies benefit more than we would expect. In other words, we might create something we call 'emergence'. According to Wikipedia, 'Emergence is conceived as a process whereby larger entities, patterns, and regularities arise through interaction among smaller or simpler entities that themselves do not exhibit such properties' (Wikipedia, 2016c). Ants provide the most outstanding example of emergence in complex systems. They are individually tiny and weak, but their colonies are huge and impressive. We see extraordinary results from these industrious insects even though they have no clear hierarchical system. No individual ant has a set of orders, only a set of rules for behavior and a purpose.

In summary, we can learn from both ants and the polar bear in order to understand what is important for increasing the probability of success for NTBFs. For everyone working with start-ups, developing companies is a fascinating mystery. The toolboxes and networks we offer help a lot, but how to make start-ups grow remains a challenge. Therefore, we need to keep searching for new and creative ways to make each company successful. Investing in new high-tech firms is therefore very similar to the ice bucket challenge: refreshing and exciting, but also very challenging.

REFERENCES

Benrath, B. (2013), 'Wir gründen in Berlin', http://www.faz.net/aktuell/beruf-chance/arbeitswelt/2500-startups-wir-gruenden-in-berlin-12119455.html (accessed 15 January 2016).

Bloomberg (2015), *BusinessWeek*, 6 January.

Bonaccorsi, A. and M. Montaina (2012), 'The public role in financing innovative companies: shifting from venture capital to seed investment', Innovation for Growth – i4g, Policy Brief No. 5.

Boyer, T. and R. Blazy (2013), 'Born to be alive? The survival of innovative and non-innovative French micro-start-ups', *Small Business Economics*, **42** (4), 669–83.

Hansen, S. (2014), *Start-up Berlin Guide* (Berlin: Berlin Projects).

Janz, C. (2014), 'Good VCs, bad VCs', http://christophjanz.blogspot.de/2014/11/good-vcs-bad-vcs.html (accessed 15 January 2016).

Ries, E. (2008), *The Lean Startup: How Today's Entrepreneurs Use Continuous Innovation to Create Radically Successful Businesses* (New York: Crown Business).

Senor, D. and S. Singer (2011), *Start-up Nation: The Story of Israel's Economic Miracle* (New York: Twelve).

Small Business Administration (SBA) (2016), 'SBIC program overview', https://www.sba.gov/content/sbic-program-overview (accessed 15 January 2016).

Wikipedia (2016a), 'Ice bucket challenge', https://en.wikipedia.org/wiki/Ice_Bucket_Challenge (accessed 15 January 2016).

Wikipedia (2016b), 'Peter Thiel', https://en.wikipedia.org/wiki/Peter_Thiel (accessed 15 January 2016).
Wikipedia (2016c), 'Emergence', https://en.wikipedia.org/wiki/Emergence (accessed 15 January 2016).
Wong, J. (2012), 'What makes the start-up system in the Bay Area around San Francisco so successful?', Lecture at High-Tech Gründerfonds Family Day.

PART 4

Innovative tools for technology transfer

9. Founding angels as an emerging angel investment model to support early stage high-tech spin-offs

Gunter Festel

9.1 BACKGROUND AND CHALLENGE

The technology transfer gap between academic research and the commercialization of scientific knowledge can be addressed by academic spin-offs that translate this knowledge into industrial applications (Grandi and Grimaldi, 2005). Spin-offs are usually more flexible and faster than established companies, given their lean structure and the absence of any prior track record (Lerner, 2005). However, there are not enough of these spin-offs in Europe, which hampers the effective and efficient commercialization of new scientific knowledge. At present, business angels (BAs) and venture capitalists (VCs) are unable to remedy this. Because BAs (and VCs) only invest in existing spin-offs, those founders who are unable to collect enough capital from 'friends, family and fools' to fund their first steps face a difficult situation. Instead of concentrating on developing their technology and finding customers, they are forced to focus tremendous efforts on raising funds.

Founding angels (FAs) can help because they provide a mechanism to support founding teams both financially and operationally from pre-spin-off phase onwards. FAs are private individuals investing own money who contribute to the commercialization of technologies originating from universities, research organizations and corporations (Festel and De Cleyn, 2013a, 2013b). By entering the spin-off development trajectory early, they are able to complement the roles of other (later) investors like BAs or VCs, as shown in Figure 9.1 (Festel and Boutellier, 2008; Festel and Kratzer, 2012). As Figure 9.2 illustrates, their contribution to the investment space is differentiated by the time of investment (before vs. after foundation) and source of investment (own vs. foreign money).

As such, FAs help to close major resource gaps in high-tech entrepreneurship, and not only in terms of finance. As an explicit part of the

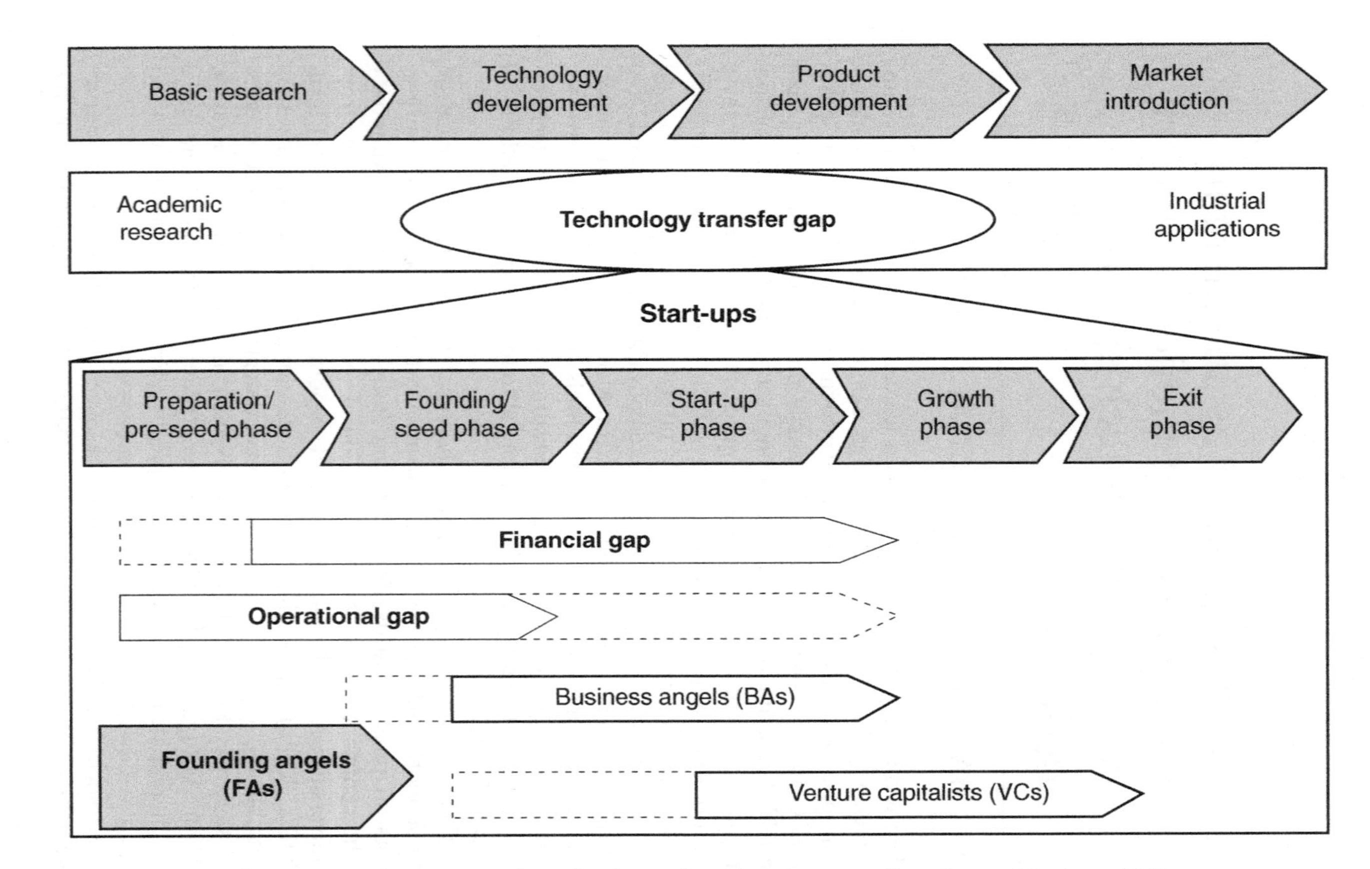

Figure 9.1 *Gaps between academic research and industrial application as well as the positioning of FAs*

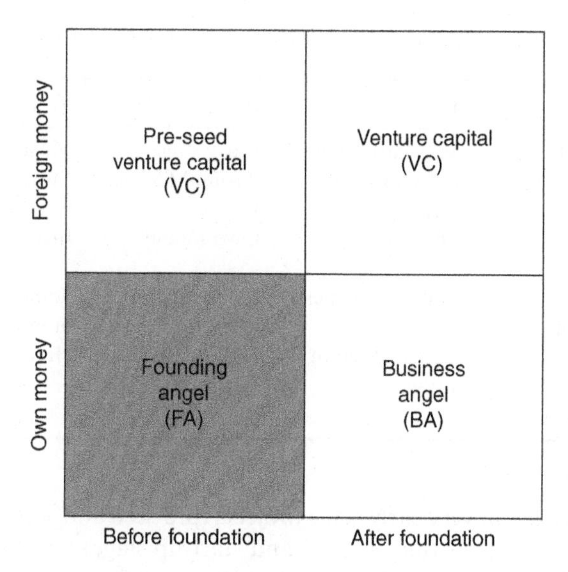

Figure 9.2 Investor categories differentiated by the time of investment (before vs. after foundation) and source of investment (own vs. foreign money)

founding team, actively involved in the earliest development stages of the spin-off, they help to bridge the gap between new knowledge coming from academic research and commercial application. They have emerged as a new type of investor, one that is significantly committed to the spin-off project both in terms of engagement (active participation as founder) and resources (financial, human and social capital and time), with the objective of achieving the successful commercialization of a new technology in exchange for shares (as opposed to consultancy fees).

9.2 CHARACTERISTICS OF FOUNDING ANGELS

There are interesting differences between FAs, on the one hand, and BAs and (pre-seed) VCs, on the other, in terms of investment volumes, exit strategy and investment holding period. This comparison is summarized in Table 9.1. There are some similarities between FAs and BAs regarding the source of the invested money, the exit strategy and the length of the investment holding period, but also significant differences regarding the investment stage, the average investment size and the intensity of the operational engagement.

Table 9.1 Comparison of FAs with BAs and (pre-seed) VCs

	FAs	BAs	VCs
Investment stage	Very early stage	Early stage	Mostly later stage
Average investment size	Around 15,000 Euros	Around 65,000 Euros	More than 500,000 Euros
Source of the money	Own money	Own money	Investors' money
Operational engagement	Some hours per day	Some hours per week	Some hours per month
Exit strategy	Less important	Less important	Highly important
Length of holding period	Long-term	Long-term	4–6 years

FAs are engaged in very early stage projects (pre-seed and seed stage), BAs in early stage projects (mostly seed and start-up stage) and VCs, increasingly, in later stage projects (mostly growth stage and only a few specialized companies in the start-up stage). Pre-seed VCs do exist and are comparable with FAs in that they primarily invest in companies co-founded with scientists and entrepreneurs, thus combining the established VC and the emerging FA investment model. But these pre-seed VCs are the exception and not the rule in the VC community.

The investment volume of FAs is between 10,000 euros and 50,000 euros per venture. This is significantly lower than BAs and other VCs whose average investment is around 65,000 euros and more than 500,000 euros, respectively. Although pre-seed VCs invest lower amounts, they are still much higher than FA investments. The reason for this is that all VCs generally have a clear incentive to increase the size of their investment per venture because their operations are financed by a management fee, typically a percentage of the funds they invest. This is totally different for FAs. They have a clear incentive to invest as little as possible in their engagements because they are investing their own money. Whereas BAs make up to four investments per year on average, FAs can handle fewer new investments because they are more operationally engaged. FAs spend between 5 and 20 hours per week supporting their engagements, and because they are typically working as individuals without any additional resources, the maximum number of investments they can handle at one time is around five.

Like BAs and VCs, FAs are also engaged with their investments up to an exit. Because they invest their own money, FAs, like BAs, are more flexible regarding the exit strategy and holding period than VCs. Prior literature indicates that informal investors, such as FAs and BAs, tend to have longer exit horizons than VCs. However, the average exit horizon of FAs is usually longer than that of BAs, because of the earlier investment stage and hence the longer development trajectory of investment projects towards potential exits. FAs own a larger proportion of the company, ranging from 10% to 50% directly after the foundation of the spin-off, compared to BAs who normally receive only a few percent for their investment. Because of the long exit horizon of an FA, both the entrepreneur and the FA have enough time to increase the value of the spin-off, which results in higher valuations when additional funding is sought from large VC funds.

The operational role FAs take within their spin-offs is, in the majority of cases, as interim chief executive officer (CEO) or chief financial officer (CFO). FAs are educated in natural sciences or engineering, plus they have a professional management career with solid financial experience. Their educational background is crucial in order for them to be accepted by university researchers who expect them to understand their science. Additional skills provided by FAs relate to their strategic and financial expertise gained in previous professional activities and to a lesser extent specific industry know-how and networks. This is very different from BAs, whose industry know-how and networks are normally seen as crucial aspects of their engagement. This indicates that strategic and financial know-how and expertise are as important for FAs as industry know-how and networks are for BAs.

9.3 FOUNDING ANGELS' ENGAGEMENT PROCESS

A process with ten partly overlapping phases has been identified as a framework for FA engagements. These phases are similar to the phases already described in the formation of spin-offs (De Cleyn and Braet, 2010) and the investment process of BAs and VCs (Mason and Harrison, 1996). Analysing the engagement process of FAs helps us to understand the specific role they play in the commercialization process of academic knowledge and to evaluate their added value. These phases are shown in Figure 9.3, while Table 9.2 illustrates the activities, decision factors and results for each process phase. The ten phases are as follows:

Project evaluation: The first phase comprises the screening and sourcing of projects. Project opportunities are evaluated to identify those with the highest potential and the best fit with the expertise of the FA. The core competence of the FA is the identification and selection of technology/market combinations with high business potential. It is important at this phase for FAs to understand value chains and the identification of

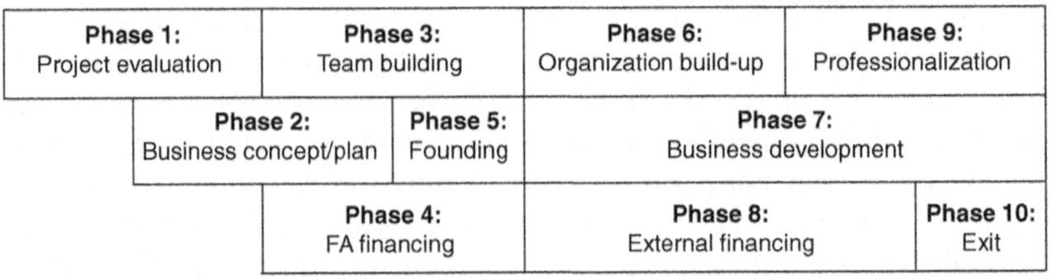

Phase 1: Project evaluation
Phase 3: Team building
Phase 6: Organization build-up
Phase 9: Professionalization
Phase 2: Business concept/plan
Phase 5: Founding
Phase 7: Business development
Phase 4: FA financing
Phase 8: External financing
Phase 10: Exit

Figure 9.3 Process phases of FA engagements

bottleneck technologies that can be further developed with their help based on a rough definition of R&D programs. Also important is the evaluation of the IP situation and how it needs to be broadened and strengthened with further development work. This phase also includes a first assessment of the team and possible skills gaps. Some FAs work with industry and technology experts to identify and pursue these new opportunities. The outcome of this phase is an attractive project opportunity for FAs if they judge the potential to be high enough and there is a good fit of topic and people.

Business concept/plan: The FA works closely with scientists from the universities or research institutions concerned to develop a business concept and, based on that, a business plan including a detailed definition of the R&D programs. The FA brings expertise and knowledge on how a business plan should look to attract investors. In some cases BAs and pre-seed VCs will support teams in writing the business plan, but this is the exception and not the rule. The FA is also responsible for the later execution of the business plan based on his/her operational role within the new company.

Team building: FAs are able to create and manage founding teams based on their experience and the credibility they have earned in other founding projects. Team assessment is a key task in this phase, notably the identification of skills gaps and the recruitment of additional team members. This enables the scientists in the team to focus on the core activities in which they excel. This clearly increases the efficiency of the team and contributes to the development of the spin-off. The FA should also be able to motivate the team and create a positive and constructive working atmosphere. A well-balanced and harmonious team is an important success factor for the spin-off.

FA financing: The FA provides the seed capital to found the company and to secure the IP position. In some cases, FAs also finance early stage research at universities or research institutions. Occasionally BAs will invest shortly after the FA's investment to increase the financial resources

Table 9.2 FA engagement process: activities, decision factors and results

	Phase	FA activities	FA decision factors to go into next phase	Result of FA activity
1	Project evaluation	First evaluation of markets and technology (incl. IP position) Rough definition of R&D programmes First assessment of the teams and missing skills	Potential high enough Fit of topic and people	Attractive project opportunities in project pipeline
2	Business concept/plan	Developing of business concept Detailed definition of R&D programmes, exit strategy and capital need Writing of business plan including detailed evaluation of markets and technology	Compelling business plan Realization probability high	Business plan to attract investors
3	Team building	Team assessment incl. identification of missing skills Recruitment of additional team members	Team completed with good 'personal chemistry'	Founding team with all skills
4	FA financing	Financing of initial activities to found the company and to secure the IP position	First money for the new company	First money for the new company
5	Founding	Evaluation of options (e.g. legal form, location) Steering of the founding process	Win-win agreement regarding distribution of shares	Start-up company
6	Organization build-up	Evaluation of needs and constraints Definition of processes and structures	Fair and sustainable allocation of work load	Efficient organization
7	Business development	Support of business development based on the industry experience and network	Positive feedback from the market (e.g. first customers)	First customers and turnover
8	External financing	Planning of investment round Search for suitable investors Negotiation of conditions/contracts	Successful investment round	Growth financing for the company
9	Professionalization	Recruitment of additional management team members	Professionalization of the management team	Complete management team
10	Exit	Search for suitable exit options Planning of the exit strategy	—	Money for founders and investors

of the spin-off. These BAs normally have a strong personal relationship with the FA. The size of this funding round is approximately 100,000 euros with only a small part coming from the FA. But the initial FA investment is crucial for the later BA engagement: without the FA, there would be no BA investment.

Founding the spin-off: When the technology is ready (e.g. proof-of-concept in the laboratory), a spin-off company is established. The FA is responsible for the evaluation of the different options (e.g. legal form, location) and the steering of the founding process. Crucial to this is a win-win agreement regarding the distribution of shares between the FA and the rest of the founding team. In most cases, an agreement with the university or research institution is signed based on exclusive rights to all relevant IP (Shane, 2002). In exchange, the university or research institution receives a pre-agreed payment and/or an equity stake in the spin-off.

Organization build-up: The FA provides intensive technical and non-technical advice to the spin-off. The new company uses the FA's seed funding to build and operate the company, typically focusing on R&D activities. The research focus is on applied research up to the development of a working prototype. The FA also helps the spin-off to obtain access to additional academic research laboratories and manufacturing facilities if required. An important consideration for the FA at this stage is a fair and sustainable allocation of workload between him/herself and the other team members.

Business development: FAs support business development based on their industry know-how and networks. This includes conducting market research, supporting product development and establishing market entry strategies. The aim is to build up a sustainable business for the spin-off through the sale or licensing of the technology or sale of services or products by acquiring cooperation partners and customers.

External financing: The FA supports the CFO of the spin-off (who is normally not very experienced at this early development stage of the company) to plan and execute the financing round. Together with the CFO, the FA is responsible for searching for suitable investors and for negotiating the terms and conditions of the investment and contract. If necessary, the FA will organize a broader investor syndicate. It is critical that the FA has a reliable network and a good reputation on the investment scene where he/she is active. Pre-set milestones are used to assess the progress of the research projects.

Professionalization of the company: In the scale-up phase, a professional management team is hired to take over responsibility from the initial management team created by the FA. Because the FA typically has an official role in the spin-off as interim CEO or CFO, this is also the phase when the FA steps back from this operational role and into an advisory role.

Executing an exit: The exit will enable the founders, FA and other investors to realize their investment in the company. Universities or research institutions will also profit if they have an equity stake in the company. The FA works with the CFO and investors to plan the exit strategy and search for suitable exit options. It is crucial in this phase that all team members and investors reach agreement on the exit strategy. In most cases, a trade sale to existing industrial cooperation partners is realized.

9.4 EXAMPLES AND INVESTMENT STRATEGY

Two examples of FA engagements in Germany and Switzerland illustrate the application of the FA concept. The first is Autodisplay Biotech in Dusseldorf, Germany (http://www.autodisplay-biotech.com), which is active in the domain of technology that enables the display of proteins on the surface of microorganisms. The technology can be applied in the field of biocatalysis, drug discovery, antibody development and bioanalytics. The FA supported the scientist who developed the autodisplay technology in securing the relevant IP rights, founding the company and attracting additional investors to enable the building of own laboratories. The additional investors requested that, in the starting phase, the FA was to be responsible for business development because of his broad network in the relevant industries. After a build-up of business development activities, the FA helped to find an external person to take over this role. Furthermore, thanks to his broad network, the FA enabled a second financing round with an industrial investor from Malaysia who invested a significant amount of money.

The second example is Butalco in Stansstad, Switzerland. This company develops new production processes for second-generation biofuels and biochemicals based on lignocellulose. The core technology is based on genetically optimized yeast and enables increased yields in bioethanol production by using C5 sugars in the fermentation process. The FA supported the scientist, a professor at the University of Frankfurt who developed tools to modify yeast, in founding the company and finding additional investors. The research was conducted at the University of Frankfurt based on research contracts that secured all the resulting IP rights for Butalco. Also, additional IP rights to broaden Butalco's technology base were acquired. Butalco was recently sold to the French company Lesaffre, enabling all shareholders to make a very profitable exit. The FA was CEO of Butalco from foundation to exit.

Both examples show that the FA investment strategy has three central features. The first is the identification of interesting markets with high potential where established companies are too slow or 'conservative'. The second is an understanding of the value chains and identification of bottleneck technologies with focused investments to develop these technologies. The third important feature is building up a strong IP position and cooperation with established companies to use their marketing and production resources and a subsequent trade sale. The uniqueness of the FA investment strategy offers clear advantages. Because the FA's engagement is at an early stage in the new spin-off, there is little competition from other investors and a huge opportunity to ensure attractive investments with high value creation potential.

9.5 CONCLUSIONS AND RECOMMENDATIONS

Founding angels, as early stage technology investors, can be defined as an investment model with huge potential to increase spin-off activities, especially at universities and research institutions. FAs are active in high-tech sectors and invest at an earlier stage of spin-off development than other types of investor. The business model combines management and capital. FAs provide business expertise and operational day-to-day support to found and develop a new company when technologies are ready to leave the laboratory. FAs are highly valuable to founders because of the time they invest in supporting them in their daily business and the vast amount of knowledge, skills and experience they bring. When comparing FAs, BAs and VCs, we see that these different investment models are in fact complementary and fit perfectly together.

By collaborating with entrepreneurs at an early stage of a company's development, FAs allow entrepreneurs to focus on the core activities in which they excel. This clearly increases the efficiency of the team and further development of the spin-off. A well-balanced and harmonious team is an important success factor for a spin-off and FAs are able to create these first-class teams based on their operational expertise and engagement. Because they work closely with founders, FAs acquire a deep knowledge of the financial situation and technological potential of the company. This means that when important decisions need to be taken, such as whether a large investment should be made, FAs can decide differently to other investors because they have a deeper understanding of and more complete information about the company.

The added value of FAs is based on (1) the time they invest in supporting scientific founders in their daily business; (2) the vast amount of knowledge, skills and experience they bring; (3) their access to relevant networks; and (4) pre-seed funding. In short, they provide a bridge between early stage scientific developments and marketable products, which is necessary due to the premature development status of university technology. Because of their experience and knowledge of a specific industry, FAs are a driving force behind the founding of new spin-offs. They keep an eye out for new scientific breakthroughs that show commercialization potential. In this way, new technologies or ideas that would otherwise remain undiscovered make it to the market. FAs have a 'pull' function in the venture business and can significantly help to close the technology transfer gap through their support of spin-off activities at a very early stage, acting as a link between academia and the business world.

REFERENCES

De Cleyn, S. and J. Braet (2010), 'The evolution of spin-off ventures: An integrated model', *International Journal of Innovation and Technology Management*, **7** (1), 53–70.

Festel, G. and R. Boutellier (2008), 'FAs as a driving force for the creation of new high-tech start-up companies', The R&D Management Conference 2008, Ottawa, Canada, 20 June.

Festel, G. and S. De Cleyn (2013a), 'Founding angels as an emerging subtype of the angel investment model in high-tech businesses', *Venture Capital: An International Journal of Entrepreneurial Finance*, **15** (3), 261–82.

Festel, G. and S. De Cleyn (2013b), 'Founding angels as an emerging investment model in high-tech areas', *The Journal of Private Equity*, **16** (4), 37–45.

Festel, G. and J. Kratzer (2012), 'The founding angels investment model – case studies from the field of nanotechnology', *Journal of Business Chemistry*, **9** (1), 19–29.

Grandi, A. and R. Grimaldi (2005), 'Academics' organizational characteristics and the generation of successful business ideas', *Journal of Business Venturing*, **20** (6), 821–45.

Lerner, J. (2005), 'The university and the start-up: Lessons from the past two decades', *Journal of Technology Transfer*, **30** (1–2), 49–56.

Mason, C. and R. Harrison (1996), 'Informal venture capital: A study of the investment process, the post-investment experience and investment performance', *Entrepreneurship and Regional Development*, **8** (2), 105–25.

Shane, S. (2002), 'Selling university technology: Patterns from MIT', *Management Science*, **48** (1), 122–37.

10. Flipping the knowledge transfer model using start-ups: how entrepreneurs can stimulate faster adoption of academic knowledge

Sven H. De Cleyn and Frank Gielen

10.1 INTRODUCTION

Research institutions, such as universities, university colleges and other public research organizations (PROs), have been engaging in knowledge transfer outside their boundaries for centuries, using mostly education and publications as their preferred channels. More recently, PROs have been trying to find other, more direct ways to inject new knowledge into applications for business and society, usually facilitated by a technology transfer office (TTO) or industrial liaison office (ILO). The TTO is usually a dedicated group of people within the PRO (sometimes as a separate department or even as a separate legal entity) with the purpose of supporting the PRO in its efforts to bring research results into (commercial) application (including but not limited to negotiating and signing licensing deals and supporting the creation of spin-off ventures). Additionally, the TTO sometimes has extra functions, including negotiating contract research agreements or some project management for externally funded research projects (including projects funded by the European Commission or local authorities). In this sense, PROs have been increasingly engaging in more entrepreneurship-related activities: establishing spin-off ventures, setting up investment funds, etc. This additional role has sometimes been described as the third mission of PROs (besides research and education) (Etzkowitz, 1998).

However, the common approach adopted by TTOs relates closely to a technology-push or inside-out approach, where new knowledge is mostly 'pushed' from the research institution towards third parties (e.g. through the sale of intellectual property, licensing or creating spin-off ventures to commercialize new technologies). This approach poses specific challenges and many PROs are struggling to derive enough benefits from their knowledge transfer activities, both in the short as well as the long term.

Additionally, small and medium-sized enterprises (SMEs) and start-ups face important challenges in gaining access to the latest knowledge, state-of-the-art technologies and research results developed at PROs (Nunes et al., 2006). This is due to the fact that SMEs manage knowledge differently from large companies (Desouza and Awazu, 2006) and PROs are not well adapted to interact with SMEs (Gibb, 2000). Additionally, many SMEs lack the 'R' in R&D (research and development). For most (Western) economies, this becomes a real issue, since SMEs are the engines of economic growth and innovation. Indeed, SMEs typically account for at least 50% of employment generated in Western economies and more than 90% of the total amount of businesses in any region (Federation of Small Businesses (FSB), 2014).

The objective of this chapter is to provide a conceptually new approach to technology and knowledge transfer from PROs (the 'flipped knowledge transfer' model, which is more demand-driven and less technology-push) and to illustrate this new approach by means of a case study with preliminary results. In this flipped knowledge transfer approach, external start-ups and SMEs play a key role and knowledge creation is no longer unidirectional, but based on co-creation. This can enable fast(er) adoption of new knowledge and technologies and inspire researchers to pursue follow-up research activities. Additionally, this model to some extent enables start-ups and SMEs to build up a real research component, as part of the entire R&D process. One of the key advantages of this new, demand-driven approach is that knowledge and research results find a more 'natural' way into market applications.

10.2 KNOWLEDGE TRANSFER AND START-UPS/ SMES

Given the increasing importance of cluster thinking and the role academic research plays in this context, knowledge and technology transfer are shown to play a prominent role in the creation of new technology-based ventures. The process of transferring knowledge and technologies from PROs to industrial actors (and broader society) has been better understood in recent years (see e.g. Inkpen and Tsang, 2005). However, current practices and academic studies have focused on the more traditional way of bringing knowledge and technologies into application, i.e. from a technology-push orientation (Siegel et al., 2003). PROs adopting this approach have traditionally been transferring new knowledge and technologies through licenses on intellectual property rights towards third parties and/or through the creation of spin-off ventures. This approach has proven successful in a number of domains, including biotechnology, pharmaceuticals and microelectronics (Zucker et al., 2002). However, in a number of other technology domains, especially those where patents are the main mechanism to protect intellectual property, the success of this strategy is less rich (Markman et al., 2005).

This technology-push approach has also proved relatively unsuccessful when it comes to start-ups and SMEs. Research capacity is one of the largest gaps separating start-ups from the marketplace — the knowledge and means to turn innovative ideas into customer-ready solutions. While larger companies are often plugged into the research community or have their own in-house research and development teams, a start-up's innovation efforts are usually driven by small teams of developers and creatives, often operating without a precisely defined roadmap. In essence, and as already stated, start-ups typically lack the 'R' part of the R&D equation.

This chapter presents the case study of a PRO in Flanders, the northern part of Belgium, which has adopted a new approach when it comes to knowledge and technology transfer activities. This approach has been coined 'flipped knowledge transfer'. In this approach, start-ups established by individuals outside the PRO play a critical role in forming the lynchpin between the entrepreneurial world and academic research. Flipped knowledge transfer is defined as follows:

> A demand-driven approach to knowledge and technology transfer activities at PROs, where the main driver to adopt new knowledge comes from start-ups external to the PRO that are actively seeking academic knowledge to reinforce their products or applications.

This flipped knowledge transfer approach can be highly relevant for a number of reasons. First, new product life cycles and technology cycles are becoming shorter (Kessler and Chakrabarti, 2003). This evolution forces companies to adopt new knowledge and technologies at a faster rate to keep up with the latest innovations and stay ahead of competition. Secondly, and related to the first reason, the half-life of knowledge is diminishing and therefore becoming obsolete (Hershock, 2011). This should stimulate companies *and* PROs to interact more frequently in order to keep up with the latest developments. Additionally, policymakers and public opinion are becoming increasingly demanding of PROs, expecting them to introduce mechanisms that ensure their research outcomes create direct added value and a (positive) impact on society and business

(Markman et al., 2005). This flipped knowledge transfer approach may provide a tool to address these needs and issues.

In the same line of reasoning, there is mismatch in time between the availability of research results and the moment when SMEs decide to build applications based on this knowledge to pursue new business opportunities (Kaufmann and Tödtling, 2002). Furthermore, the actual dissemination of research results uses channels that are difficult to access for innovation-driven enterprises. Scientific publications and conference proceedings are often locked behind payment walls and written in a language that only researchers understand and build upon. Part of this problem is solved by the open access policy of the European Union (EU) but there is still a lot of work to be done to describe research results in a way that is comprehensible for business actors and that allows them to identify future applications in their own space. The suggested approach can help bridge this time and language gap.

10.3 CASE STUDY RESULTS AND FINDINGS

This section presents the results of an in-depth case study of the flipped knowledge transfer approach. From a methodological perspective, this case study has been preceded by an elaborate literature review on the subject (a more detailed literature review paper is in the making). Using the outcomes of this literature review, the scene for this case study has been set following the guidelines of Yin (2014). Subsequently, data have been collected and coded by multiple researchers from different institutions to ensure robust data interpretations and to avoid issues with self-reported data.

We start with an in-depth presentation of the case study itself. Then, the results and outcomes are discussed at two levels: first, the main results of the case study, followed by a more general discussion of the outcomes and benefits of a more open approach to academic entrepreneurship and knowledge transfer.

10.3.1 The Case of iMinds

The subject of the case study is iMinds, a PRO in Flanders (Northern Belgium). iMinds was established in 2004 by the government of the Flemish Region under its original name of IBBT (Interdisciplinary Institute for Broadband Technology). The organization was tasked with developing demand-driven research and solutions for the digital media and ICT sector and fostering the business and societal application and adoption of newly developed technologies, knowledge, products and services. In this sense, applying knowledge and newly created technologies to address societal and business challenges has always been part of iMinds' core mission.

iMinds functions as a network integrator for research and entrepreneurship in digital media and ICT in Flanders. In this role, iMinds collaborates with universities and university colleges and other actors in the ecosystem supporting entrepreneurship (including SMEs, large companies, TTOs, incubators, pitching events, financers and others). From a research perspective, iMinds has strategic partnerships with all five universities in Flanders (Vrije Universiteit Brussel, Ghent University, Hasselt University, KU Leuven and University of Antwerp). Through these partnerships, iMinds has direct access to and relationships with the vast majority of (ICT-related) researchers in Flanders. As such, iMinds acts as a lynchpin in a Triple Helix ecosystem for the Flemish digital media and ICT community, integrating various actors and stakeholders.

The activities of iMinds are centered on two pillars: (1) collaborative and demand-driven research, in close cooperation with Flemish, Belgian and international companies, governmental agencies and other societal actors; and (2) fostering entrepreneurial behavior among researchers and external parties and supporting commercialization and other entrepreneurial activities through various programs.

In its latter role, iMinds has, since its inception, been engaging in technology and knowledge transfer activities. Whereas in the early years of the organization (2004–2011), iMinds adopted the common 'technology-push' knowledge transfer approach, in more recent years the new, flipped knowledge transfer approach has been deployed. This case study describes and analyses the potential benefits and drawbacks of this approach by examining the effects on the entrepreneurial appetite of individual researchers, the dynamics in research groups and the ability of external entrepreneurs and SMEs to gain access to and adopt new technologies and knowledge created in PROs.

10.3.2 Results of the Case Study

Since its establishment in 2004, iMinds has supported more than 90 start-up and spin-off initiatives through its incubation program (named iStart since 2011). The projects involved in this program receive pre-seed funding and intensive coaching to help the entrepreneurs make the transition from a technical proof-of-concept to a real business. The program began with a classical technology transfer approach, whereby technologies and knowledge developed by the research groups was transferred to business and society using common methods (creation of spin-offs, licensing, etc.).

However, since 2011 the program not only supports spin-off projects started by iMinds' own researchers, but also independent entrepreneurs with innovative digital media and ICT applications.

This new approach imposed a number of important challenges on the organization. In the first place, it required an additional investment in terms of funding (pre-seed funding for the incubation projects) and manpower (increasing the number of coaches and support staff). These investments are, however, relatively limited in absolute and relative terms (approximately 1 million euros per year). The second challenge related to corporate culture and mindset, especially on the research side. Opening up the incubation program to external entrepreneurs required a rapid intake and evaluation process, especially since the speed of innovation and technological progress in digital media and ICT forces entrepreneurs to keep pace with and benefit from short market windows. The overall intake and evaluation process changed from a case-by-case evaluation when a potential spin-off project applied (which took around four to six weeks from start to finish) to a call-based intake using a two-step evaluation (which only took three weeks from submission to final decision).

Since opening up the incubation program to external entrepreneurs in the summer of 2011, the number of submissions as well as the number of incubation projects supported has steadily risen (see Figure 10.1). In the period 2004–2015, more than 90 incubation projects were supported, of which 29 originated directly from academic research. These 29 projects would commonly be described as academic spin-offs (ASOs).

During the incubation program (typically lasting between six and 18 months), start-ups receive a small amount of pre-seed funding, as well as in-depth coaching. The aim of the program is to support start-up teams that have a proof-of-concept of their technology or product towards introducing their first product(s) on the market and, if necessary, towards a stage where they are mature enough to attract follow-up investments by private and public investors. The coaching program consists of daily support by a dedicated coach, complemented by workshops and one-to-one in-depth coaching by industry and subject matter experts (e.g. on marketing and branding, business-to-business sales, pitching, usability or preparing investment rounds). In some cases, the teams are reinforced by an entrepreneur-in-residence, who temporarily joins the start-up.

More importantly, since iMinds adopted its flipped knowledge transfer approach in 2011, 82 new incubation projects have received substantial support. The origin of these 82 projects is as follows (see also Figure 10.2):

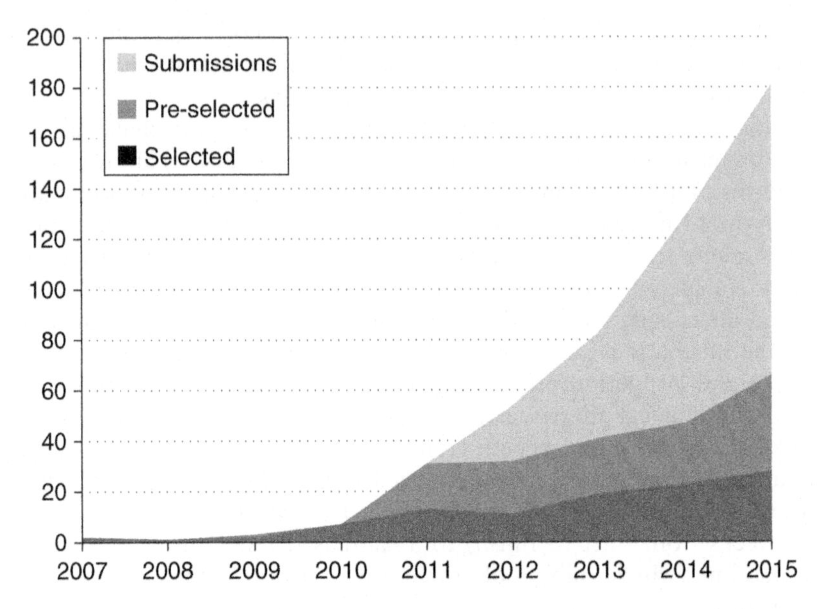

Figure 10.1 *Number of submissions and supported incubation projects (2004–2015)*

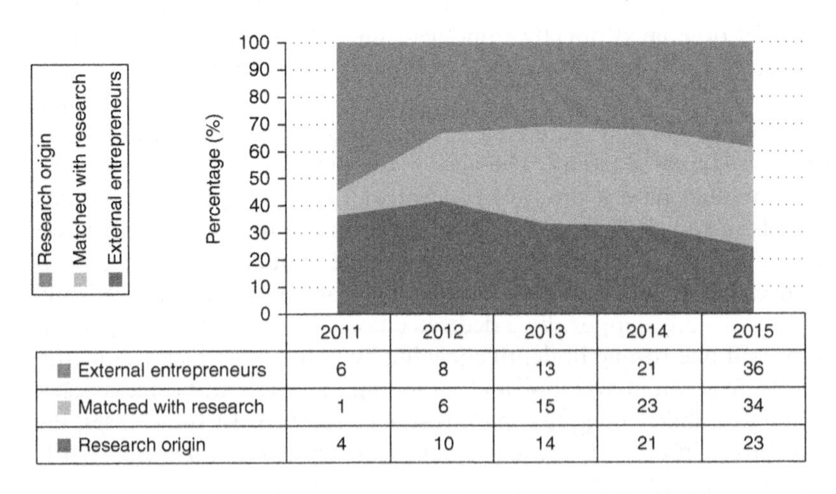

	2011	2012	2013	2014	2015
External entrepreneurs	6	8	13	21	36
Matched with research	1	6	15	23	34
Research origin	4	10	14	21	23

Figure 10.2 *Origin of iStart projects (cumulative, 2011–2015)*

- 21 projects (26%) concerned ASOs, bringing academic technologies and knowledge to the market with researchers as lead entrepreneurs;
- the remaining 61 projects (74%) concerned start-up initiatives by external entrepreneurs, where the main product idea(s) did not originate from academic research;
- moreover, 31 of these 61 projects (38% of the 'grand' total) have been matched with academic research, meaning that the external entrepreneurs have started a collaboration with one or more research groups to reinforce their technological and knowledge base and adopt and embed academic research results in their products and services.

The net result of this flipped knowledge transfer approach is quintuple.

Academic research results find a more natural way into 'real life' applications (given the demand-driven nature of the model). The start-ups benefit hugely from this direct input of (state-of-the-art) academic knowledge, while researchers are enthused by seeing their technologies 'at work'.

Researchers receive feedback from the viewpoint of implementation, which often leads to further (contract) research. In many cases, the cooperation has led to joint applications for innovation subsidies and joint research projects, to the benefit of both the start-up (providing the means to continue or even intensify the cooperation) and the research group (additional funding to ensure continuity or even growth).

Barriers preventing small, young companies and researchers from interacting have to a large extent been overcome. This not only leads to more frequent and intense cooperation between the two groups, but also creates a channel for researchers (mainly PhD students) to find good employers (and vice versa for the start-up to be able to attract highly skilled employees). This is particularly relevant for the PhD students, since 80% of them need to pursue a career outside academia after successfully defending their PhD.

SMEs and start-ups gain (easier) access to the latest developments in academic research, which enables them to differentiate their products and services through innovation, based on relevant, demand-driven research. This could give them a competitive advantage or at least the opportunity to compete with more or less equal (knowledge) weapons against larger corporations.

PROs in the long run may attract more revenue for their research activities through long-term collaboration (joint research projects, contract research, funding of PhDs, etc.). In one of the most successful cases, a start-up now funds two full-time PhD students in a research group and is involved in two joint research projects for which external funding has been obtained (one project funded by a regional innovation agency and one funded by the European Commission). This long-term cooperation has been supplemented with one-off revenue (lump sum) for the research group for the transfer of intellectual property.

10.4 IMPLICATIONS AND DISCUSSION

10.4.1 Implications of the Case Study

Research publications have long been the traditional mechanism for disseminating new knowledge and research outcomes, mainly with a view to informing and inspiring other researchers to further build on it. New methods may be needed, tailored to the needs of companies, to transfer knowledge and see it adopted. Developing a new model of thinking about knowledge dissemination is certainly an intellectually interesting exercise, where (part of) the solution may be to think out of the box and abandon a technology-push approach (or at least develop an alternative).

The implications of adopting a flipped knowledge transfer model are threefold. First, this model requires a new approach. Matchmaking between researchers and entrepreneurs needs to be done by people and organizations with enough domain expertise to identify links between the technological and knowledge needs and questions of companies (start-ups, SMEs) and research activities. This matchmaking role is not insignificant because these needs and questions are usually inexplicit and masked (entrepreneurs often do not know exactly what they are looking for). As a result, flipped knowledge transfer cannot be organized using the same, generic model that most PROs currently adopt. This new approach requires even more domain experts.

A second implication relates to the transferability of the model, which differentiates from the classical technology transfer approach. Given the nature of the industry and domain (digital media and ICT), the technologies often concern 'softer' forms of knowledge (fewer patents, more software, algorithms and methodologies). Therefore, the terminology 'knowledge transfer' seems more appropriate than 'technology transfer', even though both forms relate to transferring new technologies and various forms of knowledge. However, in the flipped knowledge transfer model, more attention is (and should be) devoted to closing the knowledge feedback loop within an ecosystem of a particular research and technology cluster. In industries with shorter R&D cycles and easier transfer of knowledge (such as digital media and ICT), adopting the flipped knowledge transfer model may happen more easily.

Placing the demand-side central is the third implication. This requires efforts from both TTOs and researchers. By positioning research groups' assets well and implementing processes that facilitate and accelerate collaboration between companies and researchers, the attractiveness of PROs towards SMEs and start-ups increases significantly. This entrepreneurial-friendly climate is a prerequisite for successfully implementing a flipped knowledge transfer model, allowing entrepreneurs to build their growth strategies based on reinforcing their products and services with research-driven innovations.

10.4.2 Discussion

Worldwide, most PROs have only been engaging explicitly in technology and knowledge transfer activities during the last few decades. Compared to conducting research and providing education, which have been part of PROs' core activities for centuries, this relatively new activity domain is subject to debate and experimentation in order to find the most appropriate model to achieve success. However, many PROs have not yet found the best recipe to deploy (highly) successful technology transfer activities.

The value of this case study lies mainly in highlighting a different approach. The traditional technology transfer approach taken by PROs builds on a technology-push or inside-out model, where new technologies developed within PROs are 'pushed' towards business and society where they will hopefully find applications and be adopted. Prior studies have, however, demonstrated the relative prematurity of most PRO technologies when brought to the market (Shane, 2004), requiring substantial additional efforts in translating the technology into market-ready products and applications.

The iMinds case study proposes a different, more outside-in approach. The so-called flipped knowledge transfer model adopts an inverse perspective, whereby market players interact much more frequently with researchers and whereby PRO technology and knowledge is 'pulled' outside the PRO in a more demand-driven way. In this model, start-ups and entrepreneurs play a pivotal role.

Even though this model is relatively new, the first results indicate interesting potential for PRO technology transfer. They indicate a positive impact for individuals (mainly researchers, who can find real-life applications for their knowledge and interesting career opportunities), start-ups and companies (who can gain a competitive advantage and improve their products with state-of-the-art research outcomes, as well as access to highly-skilled people) and PROs (who receive industry input, and potentially additional revenue streams and more extensive cooperation with the business world). However, the model and its effects, both directly and indirectly as well as in the short and long term, deserve further attention and research. It opens the debate for more studies on the impact of technology and knowledge transfer programs, as well as continuing the search for best practices and alternatives to boost the commercialization of academic research output.

A second important contribution of this case study is its value in making explicit the potential benefits for PROs of opening up their entrepreneurship programs and support towards 'externals'. Tearing down the walls of the academic 'ivory tower' may have a substantial positive impact both on organizations and individuals. It is clear that not all PROs can open up their TTO activities and entrepreneurship programs as widely as iMinds has. It may be difficult for universities or other PROs to spend scarce resources on supporting totally unrelated start-ups or SMEs. Opening up such programs has far-reaching implications for an organization, both in terms of competences required and resources and processes that need to be adapted (mainly, but not limited to TTOs). However, the authors firmly believe that a certain level of openness, embracing at least those entrepreneurs and start-ups that can be directly linked to researchers and benefit from academic knowledge, has a positive effect on the start-up, the PRO and the individuals involved (thereby only excluding the category of fully external entrepreneurs in Figure 10.2). More frequent interaction between researchers and entrepreneurs fosters knowledge spillovers, reinforces the economic tissue of a region and acts as an enabler to form a cluster of innovation.

The importance of adopting such a flipped knowledge transfer model from a societal point of view touches a number of aspects. First and foremost, it reinforces the knowledge base (and often also the technological base) of start-ups and SMEs. This in turn could create competitive advantage and strengthen their long-term market position. In fact, this flipped knowledge transfer creates (external) research capacity, an asset that most (small) companies cannot afford to invest in alone. Secondly, adopting this model enables regions or countries to make more efficient use of scarce resources invested in public R&D. Using the more demand-driven approach (in conjunction with the classic operational mode of PROs), new knowledge finds its way into applications at a faster rate and (usually) creates more benefits for local companies, and not only for larger (and often more international) corporations. The third main benefit for society lies in the fact that researchers on one hand and entrepreneurs and employees in start-ups and SMEs on the other improve their human capital (skills, experiences and expertise). This human capital can then be deployed in other settings, projects or forms of cooperation.

PROs may experience a number of obstacles when trying to implement a flipped knowledge transfer model. The first may relate to the legal framework. Not all PROs will be allowed to spend part of their scarce resources on supporting (initially) unrelated companies. Secondly, this transition requires a new set of procedures and processes to enable PROs to manage such cooperation in a flexible and speedy manner. Specifically for start-ups, the speed of go-to-market and agility are critical for their chances of survival. For this kind of cooperation to be successful, both sides need to adapt. Another important element relates to resources in terms of people, expertise and (eventually) funding. A flipped knowledge transfer approach to fostering entrepreneurship around a PRO can only be successful if it becomes part of a PRO's strategy. The authors believe, based on the first results, that sector expertise is required to fully realize the approach's potential, which has implications on a TTO's staffing requirements. A fourth and last major factor for the successful implementation of the flipped knowledge transfer model relates to the demand-driven nature of PRO research. iMinds is by definition a demand-driven PRO, where (almost) all research efforts are conducted in close cooperation with industry and society stakeholders. As a result, research efforts are closer to end-user applications and to the market. This may impose a challenge for a number of PROs, especially in safeguarding a good balance between fundamental and applied research. However, based on the results of this case study, the authors believe that even when research is partially reoriented towards more demand-driven needs, this will in the long term also positively impact fundamental research, since new challenges and opportunities can be identified based on the concrete implementation of new knowledge and technologies in entrepreneurial settings.

This case study has explored the topic of a flipped knowledge transfer model as a tool to stimulate the creation of clusters of innovation and to strengthen market adoption of academic research in concrete applications. The more demand-driven approach stimulates start-ups and SMEs, which are key in creating clusters of innovation in a region and are responsible for the majority of wealth and job creation. In this sense, PROs can play a crucial role at regional level in reinforcing, and also stimulating, entrepreneurship. Despite the limitations of this case study, the first results of adopting a flipped knowledge transfer model have proven positive for the PRO, researchers and start-ups involved. PROs and their TTOs will play an increasingly important role in regional dynamics, especially for technology clusters and start-ups. This flipped knowledge transfer model can be an additional tool in truly realizing the potential of the Triple Helix Model and creating innovation hubs where new start-ups flourish and safeguard (or even improve) a region's wealth.

REFERENCES

Desouza, K.C. and Y. Awazu (2006), 'Knowledge Management at SMEs: Five Peculiarities', *Journal of Knowledge Management*, **10** (1), 32–43.

Etzkowitz, H. (1998), 'The Norms of Entrepreneurial Science: Cognitive Effects of the New University – Industry Linkages', *Research Policy*, **27** (8), 823–33.

Federation of Small Business (FSB) (2014), *Statistics. Federation of Small Businesses*, http://www.fsb.org.uk/stats (accessed 15 January 2016).

Gibb, A.A. (2000), 'SME Policy, Academic Research and the Growth of Ignorance, Mythical Concepts, Myths, Assumptions, Rituals and Confusions', *International Small Business Journal*, **18** (3), 13–35.

Hershock, P. (2011), 'Information and Innovation in a Global Knowledge Society: Implications for Higher Education', in D.E. Neubauer, *The Emergent Knowledge Society and the Future of Higher Education: Asian Perspectives* (London: Routledge), 12–48.

Inkpen, A.C. and E.W.K. Tsang (2005), 'Social Capital, Networks, and Knowledge Transfer', *Academy of Management Review*, **30** (1), 146–65.

Kaufmann, A. and F. Tödtling (2002), 'How Effective is Innovation Support for SMEs? An Analysis of the Region of Upper Austria', *Technovation*, **22** (3), 147–59.

Kessler, E.H. and A.K. Chakrabarti (2003), 'Speeding Up the Pace of New Product Development', *Journal of Product Innovation Management*, **16** (3), 231–47.

Markman, G.D., P.H. Phan, D.B. Balkin and P.T. Gianodis (2005), 'Entrepreneurship and University-based Technology Transfer', *Journal of Business Venturing*, **20** (2), 241–63.

Nunes, M.B., F. Annansingh, B. Eaglestone and R. Wakefield (2006), 'Knowledge Management Issues in Knowledge-intensive SMEs', *Journal of Documentation*, **62** (1), 101–19.

Shane, S. (2004), *Academic Entrepreneurship – University Spinoffs and Wealth Creation* (Cheltenham: Edward Elgar).

Siegel, D.S., D. Waldman and A.N. Link (2003), 'Assessing the Impact of Organizational Practices on the Productivity of University Technology Transfer Offices: An Exploratory Study', *Research Policy*, **32** (1), 27–48.

Yin, R.K. (2014), *Case Study Research: Design and Methods* (Thousand Oaks: Sage Publications).

Zucker, L.G., M.R. Darby and J.S. Armstrong (2002), 'Commercializing Knowledge: University Science, Knowledge Capture, and Firm Performance in Biotechnology', *Management Science*, **48** (1), 138–53.

11. Stimulating student entrepreneurship within a traditional university model: the case of the AU Student Incubator

Mia L. Justesen, Rajiv V. Basaiawmoit, Flemming K. Fink and Kirstine V. Moltzen

11.1 INTRODUCTION

Universities in general are very conservative institutions with proud academic traditions. Several ranking systems attempt to define what it means to be among the best universities in the world. Criteria taken into consideration include research activity, research funding, number of peer-reviewed scientific publications (in ranked journals), number of citations, number of Nobel Prize winners, reputation of educational programs, number of international students, etc.

These criteria are all defined by the academic world. However, increasing demand from society and industry urges universities to become better at bringing research-based knowledge into society and into application in the real world. Research outcomes from a joint research project funded by a major industrial partner will find their way to real-world applications. Research findings are also embedded in educational programs at different levels – from short courses to entire degree programs. Graduates are thus expected to both learn and transfer knowledge when they take up jobs in industry or government. However, this process is not only time-consuming but also graduates often find it difficult to apply their university-acquired knowledge practically. Thus, entrepreneurship is largely seen as a route to fast track academic knowledge to real-world application.

11.1.1 The Entrepreneurial University

In 2008, the president of Aarhus University (AU) decided that the university should take new initiatives and be more proactive in sharing knowledge to benefit society. These new initiatives included setting up a student incubator and developing support concepts for student start-ups, and integrating more entrepreneurship courses into different degree programs. In 2010, AU received the award of 'The Entrepreneurial University' in Denmark for its strategic plan to develop and combine university praxis with entrepreneurial educational activities. This award included major funding for activities during three years. A small part of this funding was reserved for further research activity on entrepreneurship teaching and learning. The main focus was on entrepreneurship praxis. The AU Centre for Entrepreneurship and Innovation was made responsible for assisting university departments with course development and implementation and for developing Student Incubator activities.

11.1.2 Student Incubator

The Student Incubator (SVAA) was established in a separate building offering adequate space for a range of activities. One large room was equipped with office furniture where students could work on their start-up concept in close collaboration with other students and supervised by SVAA staff. Smaller rooms were designated for meetings between students and potential partners and investors. A workshop area, equipped with adaptable furniture and technologies, was also created for group sessions and guest presentations by experienced entrepreneurs and other experts.

The set-up was flexible and dynamic. More than 40,000 students are enrolled at AU, and in its early days the incubator was also open to students from university colleges in the city of Aarhus. This is no longer the case because most colleges have now started creating their own incubators inspired by the experience of SVAA. Currently, SVAA daily hosts more than 100 students, all from AU. The starting point for SVAA was to focus on the 'learning' context in which a student finds him/herself at university, i.e. helping the student to gain new 'entrepreneurial competences' and also to understand via practice what it takes to become an entrepreneur. Thus, at a very early stage SVAA described its mission as follows: '*We develop the individual to qualify the individual to develop his/her company*'.

11.2 WHY FOCUS ON STUDENT ENTREPRENEURSHIP?

Entrepreneurship is currently attracting interest from all sectors and stakeholders – from governments to industry, from job seekers to employers and, most importantly, from students of various ages and disciplines.

Once perceived as purely the domain of Silicon Valley, entrepreneurship now captures the imagination of citizens, governments and businesses alike the world over. In parallel, there has been an increase in research on entrepreneurship, entrepreneurship education, start-up and small business clusters. This increased awareness, coupled with an upsurge of media coverage and the rise of disruptive (to business-as-usual) crowd-funding platforms such as Kickstarter, Indiegogo, etc. – seen as alternative and relatively easy ways to finance start-up projects – are probably all contributing factors to entrepreneurship being more mainstream than ever before.

Against this backdrop, student entrepreneurship has also prospered. But until now it has operated under the radar since current infrastructures within student-based ecosystems such as schools, colleges and universities are only beginning to explore its potential. In universities, for instance, entrepreneurship has largely remained within the purview of employees and to a certain extent PhD students, support for which is provided by university management and institutions such as the Technology Transfer Office. As the name of the latter suggests, most of these activities are aimed at technology commercialization, which is only a fraction of what entrepreneurship actually represents. In addition, most graduate and undergraduate students typically fall outside the classic technology transfer approach.

This raises an important question: Why focus on student entrepreneurship? And a subsequent question: What is the importance of student spin-off companies?

Within the university context, we can look at these questions from three perspectives: that of the student, the university (as in management) and the government (in terms of policy).

11.2.1 The Student Perspective

(1) It is often said that entrepreneurs should fail fast and early and learn from their mistakes. Failing within the safe confines of a university while pursuing an education can be a healthy way to reduce the 'fear of failure' among students.

(2) The rise in popularity of entrepreneurship is accompanied by media glorification of 'hero' entrepreneurs (such as Richard Branson or Steve Jobs) to such a level that they become inaccessible and students are unable to relate to these success stories. While they may 'worship' such 'heroes' or be inspired by them, when asked if they could imagine themselves reaching such heights, students often reply in the negative. Introducing them to successful entrepreneurs who are students or recent graduates narrows this gap and can act as an active enabler of student entrepreneurship.

(3) Irrespective of start-up success, the fact that students go through the entrepreneurial process makes them more self-aware. They can use this knowledge at different stages of their life, as intrapreneurs (entrepreneur within a firm) in their jobs or as entrepreneurial agents in their non-start-up lives whether in academia or government.
(4) Exploring what it means to be an entrepreneur now can be of use in the future.
(5) The employability rate among student entrepreneurs is high. Established companies are looking for graduates with entrepreneurial skills and academic knowledge to increase their innovation activities and gain competitive advantage.
(6) Using their university-gained knowledge in a start-up means that student entrepreneurs learn how to apply their knowledge and create value in practice. This increases their motivation to study and also enhances personal awareness of their competences and talents.

11.2.2 The University Perspective

(1) Student start-ups proclaim that the university from which they start is trendy, relevant, up-to-date and competitive (i.e. moving with the times rather than being rooted-in-the-past).
(2) Start-up programs attract talented students, creating a pool of future investors, mentors and donors that could strengthen national and international networks and reputation.
(3) Student start-ups commercialize knowledge that may otherwise go undeveloped within the university. Transforming 'in house' knowledge into sellable goods is expensive and uncertain. Most universities do not have the skills, will power, discipline, financial resources, space or networks to do this. In many cases, student spin-off companies are necessary to transform academic knowledge into market offerings, attract capital and validate customer value.
(4) Student spin-offs help universities accomplish their core missions of research, teaching and community development. They provide faculty with knowledge that is useful for educating students, and they increase awareness of the practical value of undertaking university research.
(5) Working in a student incubator gives the student another perspective on his/her academic expertise: 'I know something more than students from other academic environments'. The experience of the AU Student Incubator is that immersion in such an interdisciplinary environment reinforces the student's academic identity and increases his/her motivation to learn. Consequently, the university produces

talented students with high-quality academic and entrepreneurial skills that meet demand from society.

11.2.3 The Government Perspective

(1) Student start-ups contribute to the economic development of the region where the university is located. They generate jobs (including jobs for students and knowledge-intensive jobs), diversify the local economy, satisfy customer needs and attract talent and investment.
(2) Student entrepreneurship increases the return on government investment in university R&D, an issue of increasing concern among policymakers and taxpayers. Michelacci (2003) has shown that when the stock of knowledge is high and the amount of entrepreneurial skills is low, an increase in R&D reduces economic growth. When entrepreneurial skills at the university are low, returns on large R&D investments are also low. In addition to being knowledge-transfer mechanisms, student spin-offs increase the level of entrepreneurial activity at a university, which in turn increases the university's return on its R&D (Bailetti and Doerr, 2011).

These three different perspectives highlight the need for a concerted effort to drive and support student entrepreneurship. AU recognized this in the early days of entrepreneurship growth and set in process the infrastructural investments necessary, both in terms of Entrepreneurship Education (EEd) and the physical infrastructure of SVAA, coupled with a strategy focused on individual development. The next section of this chapter will focus on SVAA, after which we will cover the second infrastructural investment in entrepreneurship education and discuss how these two combine to make the AU model unique in the domain of student entrepreneurship especially in Northern Europe.

11.2.4 Definitions

Before we proceed with the SVAA case study, we will highlight the definitions of terms and concepts we use.

- *Entrepreneur*: We have primarily used the definition of nascent entrepreneur by Davidsson (2006) as one who initiates serious activities or is actively involved in the creation of a start-up. However, we realize that there is a continuum in the types of students at a university who approach entrepreneurship. These range from the curious, to those with some entrepreneurship intent, to those who already see themselves as entrepreneurs and are actively engaged. We therefore use two other definitions in addition to nascent.

- *Entrepreneurial curiosity*: For those who are curious, which, based on our experience at SVAA, is a relatively large component. These students are curious (from slight to very) about entrepreneurship in general and what it entails but also curious about information that could motivate them towards entrepreneurship (which is thus different from simple or general curiosity) (Jeraj and Antoncic, 2013).
- *Entrepreneurial intent*: For students coming to SVAA with a conviction that starting a company might be a suitable option for them (Davidsson, 1995).
- *Intrapreneur*: We also encounter and encourage the development of intrapreneurs – defined as entrepreneurs, or working entrepreneurially, within a pre-existing organization.
- *Start-up*: The definition we use of start-up is that of an early-stage company or organization that is searching for a scalable and repeatable business model (Blank and Dorf, 2012).
- *Individual development*: We work with the concept of individual development by focusing on developing the entrepreneurs so that they can further develop their entrepreneurial opportunity. Thus, the definition encompasses our focus on personal development improving entrepreneurial self-efficacy in an iterative process throughout one's entrepreneurial journey irrespective of the length of that journey.

11.3 THE SVAA (STUDENT INCUBATOR AARHUS UNIVERSITY) EXPERIENCE

Empowering students with entrepreneurship skills and creating an entrepreneurial ecosystem in which students can safely learn 'about', 'through' and more importantly 'for/in' entrepreneurship (Heinonen and Hytti, 2010) has always been the overarching goal of SVAA. The incubator has gone through multiple stages of evolution based on both planned monitoring and evaluation approaches and responsiveness to feedback from the students but also 'market' changes and evolution of the environment in which we operate. This represents a flexible operating structure that is based on some concrete principles but reacts to the fast-changing market around it. However, instead of documenting the evolution of SVAA we have decided to highlight the current and most updated state (at the time of writing) of the incubator.

The overall strategy of SVAA is to provide fertile ground for an entrepreneurial culture to take root within the university setting and then spread this across the university. This happens in parallel with the establishment of an infrastructure that supports and drives a thriving entrepreneurial ecosystem. One of the primary focus points in SVAA is individual development so that students can develop entrepreneurial skills using their academic abilities. This then translates into start-ups, intrapreneurs or individuals with enhanced entrepreneurial mindsets.

The activities we undertake are aimed at establishing the aforementioned entrepreneurial culture that supports individual development and using feedback and insights from these activities to design strategies and programs that support the overall strategy of simultaneous culture and ecosystem development. We therefore divide this section into two main sub-sections that highlight the core objectives of SVAA: (1) individual development; and (2) program design.

11.3.1 Individual Development

We continuously work on developing students' competences in three specific areas:

- the person/team
- the idea/product
- the business

We meet students at each of the aforementioned stages. These multiple touch points are critical in the strategic direction taken by SVAA towards establishing a broad-based entrepreneurial culture with easily accessible entry points. They also underline our focus on individual development and our ability to meet students at key moments of their entrepreneurial journey.

That said, all students (be they entrepreneurially inclined, curious or active) are interviewed at the beginning to gauge their understanding of their abilities through a mapping of academic and personal competences. Regardless of whether or not the students have an idea when we meet them, we have programs supporting both starting points. This will be elaborated below. A mapping of business knowledge is added if required, depending on the individual and concept.

Connecting this to our theoretical starting point, we see entrepreneurship as a discipline where personal and social competences such as creativity, openness and the ability to act are developed. At SVAA these competences are stimulated through an evidence-based approach grounded in Effectuation Logic (Sarasvathy, 2009) and the push-method

(Kirketerp, 2011). Effectuation Logic focuses on the next, best action; by taking small, fully controlled steps, enterprising behavior becomes manageable and possible. The push-method describes the importance of pushing theory towards action in order to internalize enterprising behavior and – vice versa – pushing action via reflection so that action is consciously taken and, therefore, can be improved.

At SVAA we perform this interplay between action and reflection through one-to-one sessions with students, sessions between groups of students, and joint workshop sessions. Furthermore, we also apply specific tools such as the Business Model Canvas (Osterwalder and Pigneur, 2010) and Learning Logs that improve entrepreneurial self-efficacy and independence of the student/entrepreneurial team.

One thing that probably distinguishes SVAA from other student incubators is our recognition of the fact that a student contacting us is already at some stage of an entrepreneurial journey. Structuring this into our program design (multiple entry points) allows us to offer students an entry point that acknowledges the stage they have reached on their entrepreneurial journey. Our understanding of the entrepreneurial journey (which is both evidence- and experience-based) is shown in Figure 11.1, which depicts the process and entry points to which students have access. This flowchart serves as a precursor to our program design (Figure 11.2) and also emphasizes how our focus on individual development feeds into the program design strategy.

A student's first-time encounter with SVAA generally happens at events, fairs, via marketing materials or through other students. As Figure 11.1 shows, contact with the student can end at this first encounter. Either the student does not contact SVAA again (this does not necessarily translate into an end of his/her entrepreneurial ambitions because he/she may be considering other options/paths not offered by SVAA), or the student prefers to attend a one-off Entrepreneurial Awareness Activity (EAA) at the incubator. These recurrent EAA events also aim to orient the student back to SVAA for a further, more serious, deliberation on his/her entrepreneurial journey.

The next phase of the journey normally entails the development of initial entrepreneurial intent via a series of clarification workshops (bearing in mind that we offer multiple entry points for students to choose from). After these workshops, students can again opt out of their entrepreneurial journey. The reasons students generally give for opting out at this stage include: 'the idea is not feasible'; 'entrepreneurship is not for me *right now*' (citing study or other concerns); or 'entrepreneurship *in general* is not for me'. Again, the fact that they opt out here does not necessarily translate into the end of their journey, because we have seen cases of

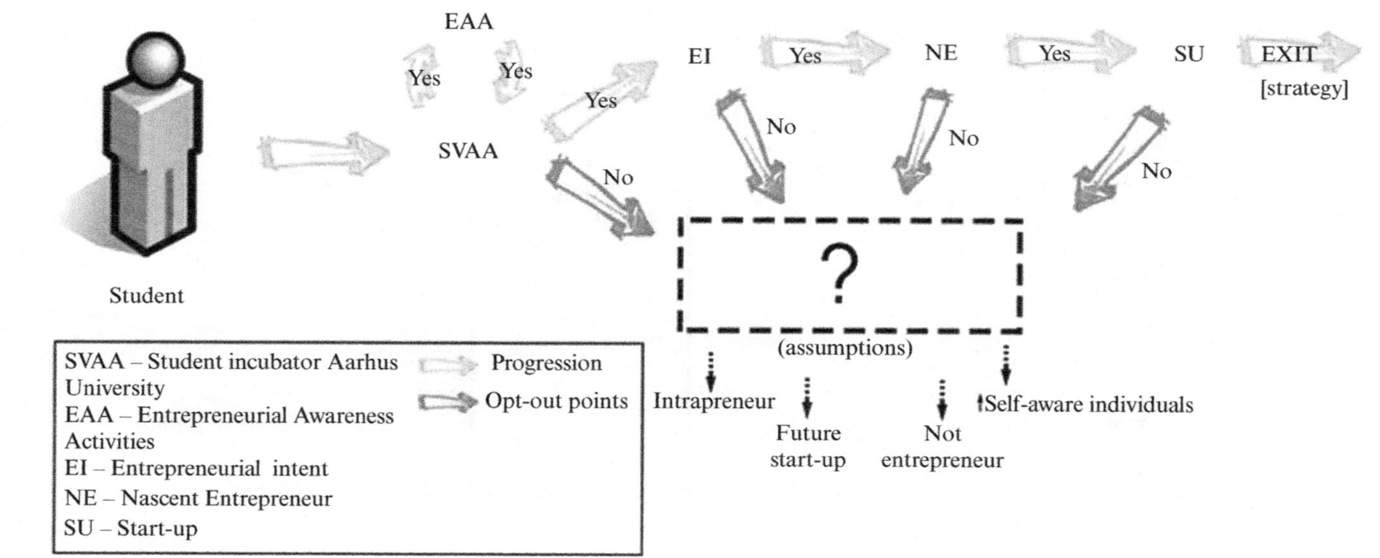

Figure 11.1 *Flowchart of a student's entrepreneurial journey and the choices a student makes while interacting with SVAA*

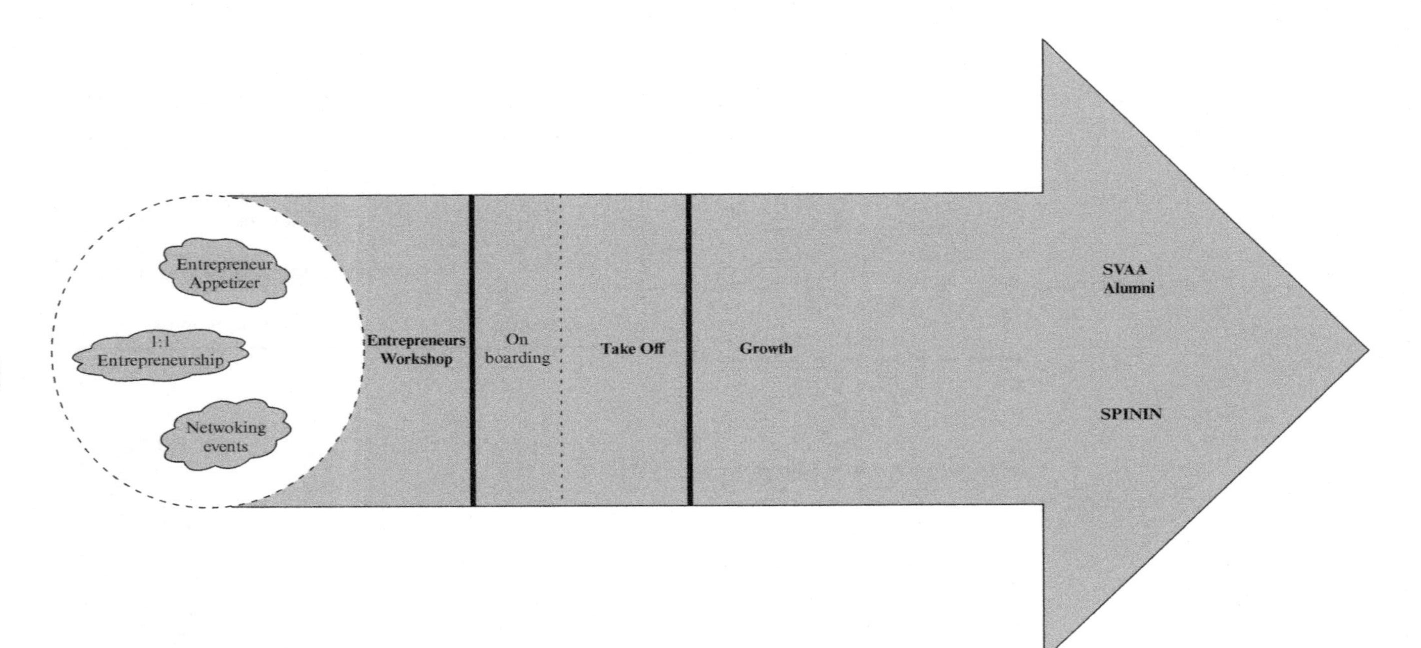

Figure 11.2 SVAA's program design

students returning to SVAA after a break. In addition, they can also apply the entrepreneurial learning they have gained thus far in their lives and/or become intrapreneurs.

For those who continue, i.e. those who still find entrepreneurship interesting and who have a concept (for a start-up), we support their further development into nascent entrepreneurs and offer them an appropriate program (see more in section 11.3.2 below). Again, just as in earlier stages, students can still opt out of the entrepreneurial activity. They will already have gained significant knowledge about entrepreneurship and this may help them to start up something later in life or become intrapreneurs. However, these remain assumptions and have not yet been investigated. As Figure 11.1 shows, our assumptions include future start-ups, intrapreneurs, individuals with higher self-awareness and, of course, those who opt out of entrepreneurship altogether with no intention of starting up a business in the future.

Successful completion of this phase is the setting-up and running of a start-up (these entrepreneurs are allowed to remain in SVAA for as long as they are students and developing their start-up). At this stage they also plan an 'exit strategy' from SVAA. Potential exit strategies include plans for new entrepreneurial environments (i.e. moving to other environments), creating advisory boards and formal organization structures (i.e. legal creation of their start-up), or joining other state-sponsored infrastructures.

Furthermore, we wish to underline that many students already are – or know that they want to become – a nascent entrepreneur when we meet them for the first time. These students are not required to go through all the preliminary steps mentioned above. We meet them at the stage they are at and provide the program that best suits their needs (see entry points in Figure 11.2).

11.3.2 Program Design

The aforementioned process (Figure 11.1) indicates an entrepreneurial journey that is both evidence- and practice-based and in which 'individual development' is a guiding theme. This approach, coupled with the multiple entry point logic, feeds into SVAA's overall strategic program design. The result is a progression-based framework that is flexible enough to accommodate students' diverse entrepreneurial ambitions and indicate the degree of involvement required of them. This framework is illustrated in Figure 11.2.

11.3.2.1 Explanation of each stage

- *Stage 1* – The keyword defining the first stage is 'Inspiration'. Here the student is introduced to entrepreneurship during a one-off meeting and no specific demands are made. There are three introductory programs: *Networking Events*, *Entrepreneurship 1:1* and *Entrepreneur Appetizer*.[1]
- *Stage 2* – The second stage focuses on 'Clarification' and consists of one five-part program, the *Entrepreneurs Workshop*. In general, students entering the *Entrepreneurs Workshop* already have some degree of entrepreneurial intent and a business idea they want to test during the program.
- *Stage 3* – In the third stage, we 'Develop Business' and the students work towards developing their concept into a start-up in the *Take Off* program. Here we work with nascent entrepreneurs and full-blown entrepreneurs and focus on establishing financially sustainable start-ups.
- *Stage 4* – In the final stage, start-ups with 'Growth Potential' are extensively supported to become high-growth companies via the *Growth* program.

Finally, as Figure 11.2 shows, nascent and full-blown entrepreneurs are enrolled in the SVAA Alumni corpus when they leave the incubator.

11.3.2.2 Considerations

- *Theoretical* – In terms of evidence-based design it is clear that the first two stages focus primarily on screening for entrepreneurial intent, whereas the two latter stages develop nascent and full-blown entrepreneurs. Stage 1 also contributes to the exploration and development of entrepreneurial curiosity. All of these are well-established theoretical constructs that we also observe in practice at SVAA.
- *Practical* – In terms of experience-based design, the above framework also shows how it is linked to the entrepreneurial journey that a student experiences. Furthermore, even though a clear progression is visible in the four stages, it is possible for the student to choose a program best suited to his/her situation at any given time. We will exemplify this below in short descriptions of some of our student entrepreneurs' starting points and journeys.
- *Infrastructure* – SVAA consists of an open office space with 23 desks, 15 smaller offices, five meeting rooms, and a workshop area that can accommodate up to 50 people. The *Networking Events*, *Entrepreneur*

Appetizer and *Entrepreneurs Workshop* are all conducted in the workshop area whereas students who are developing start-ups primarily occupy the office spaces and meeting rooms.

11.3.2.3 The programs

We will now briefly describe the specific programs involved in each stage.

Stage 1: Inspiration

Anita Dalsgaard, who studied nutrition, is an example of a student who was only mildly curious when she visited SVAA with fellow students – and one year later published her first cookbook, Enkelt, *after attending* Take Off. *She was motivated by the entrepreneurial culture and received the push she needed to realize her idea. One of Anita's greatest strengths has been her ability to involve her network in the process.*

Networking Events are morning meetings that take place every two weeks at SVAA. The format is simple: breakfast, a short presentation on an entrepreneurial topic, and networking for those who have time to stay. These events are open to students and non-students in order to boost the networking element. *Open Events* allow our partners or entrepreneurs themselves to host events sponsored by SVAA. Again, these events are open to students and non-students and the objective is to give them a taste of the SVAA entrepreneurial environment that will spark entrepreneurial intent.

The aim of *Entrepreneurship 1:1* is to provide students with easy access to information about entrepreneurship and SVAA programs. Students can send an email or call a SVAA consultant to request information about incubator activities or courses. If a student already has a well-defined idea and seeks specific business advice he/she is offered a meeting with a consultant. The result of this meeting can be that the student has sufficient knowledge to continue independently or becomes interested in joining one of the SVAA programs.

Entrepreneur Appetizer targets students who find entrepreneurship interesting but who do not yet have a business idea. The program is still being developed based on the lessons learned from a previous SVAA program, *Sand Box*. The aim of *Sand Box* was to enable students to develop ideas independently based on their academic competences as well as their personal preferences and interests. The same will apply to *Entrepreneur Appetizer*; however, the demand for entrepreneurial support has changed in recent years: more students have preliminary ideas and are not afraid to share them when we meet them for the first time. Furthermore, today our EEd covers most *Sand Box* elements. Therefore, *Entrepreneur Appetizer* will be a smaller program requiring fewer resources. The main purpose is to stimulate curiosity, form a network within the entrepreneurial field and teach the students how to develop an entrepreneurial mindset and thereby create entrepreneurial opportunities. We envisage the program consisting of one or two workshops.

Stage 2: Clarification

Jeppe Ibsen, studying a Master of Engineering, attended the Entrepreneurs Workshop *in 2012. Jeppe wanted to revolutionize the way we use 3D printers; however, he only thought of this as a hobby until he entered the program. The* Entrepreneurs Workshop *gave Jeppe another perspective and now he is determined to find a suitable business model. Jeppe is now part of the* Take Off *program where he is working on his start-up, Ragnar Extruder, together with his business partner, Sylvester.*

This stage targets students who are already interested in entrepreneurship and seek clarification regarding their entrepreneurial endeavors. The *Entrepreneurs Workshop* is designed to further develop their entrepreneurial intent. The program consists of five workshops (of three hours) spread over a five-week period. Students entering the *Entrepreneurs Workshop* already have a business idea. They want to develop that idea into a business concept and explore whether there is potential for creating a start-up.

The program is designed to accommodate a large group of participants and roughly 40 students participate in each edition. The program facilitators focus on creating a collaborative culture where ideas and thoughts are openly shared. As well as working on their ideas, students build a network of peers with similar interests and start to form an entrepreneurial identity.

During the program, it is made explicit that entrepreneurship is a mindset that can be trained and mastered. Participants often alter their ideas during the program, for example, based on feedback or because they combine their ideas with those of others. Furthermore, during the program the facilitator is free to adjust content to meet the needs of specific participants. It is important to note that SVAA is a practice lab for business ideas and not an academic exercise.

Stage 3: Develop Business

Peter Kenney, studying medicine, is developing a nose filter for people suffering from allergic reactions during the pollen season. He entered the Growth *program with a fully developed business idea. Therefore, he did not need the same type of assistance as other nascent entrepreneurs. Rather, he has benefited from insights into how to manage the mechanisms associated with involving external stakeholders and investors and how to define his own role in his start-up.*

By this stage, all participating students are now engaged in developing a start-up. There is no time limit to the *Take Off* program, as long as the business is still developing and the entrepreneur is studying at AU. Students must apply to join the *Take Off* program, and there are four application rounds per year. Those accepted are then introduced to SVAA, our methods and other entrepreneurs in three 'Lead-In' workshops.

Take Off is individually tailored to each entrepreneur's needs (for example, business consultancy, coaching and workshops on demand). However, some activities are mandatory: entrepreneurs must meet an SVAA consultant at least once every three months. The purpose of these meetings is to review/adjust action plans, reflect on the entrepreneur's learning process and discuss the development of the business model. Furthermore, entrepreneurs are expected to meet with their peers (another start-up in SVAA) once a month to share experiences.

Stage 4: Growth

Morten Fryland and Lasse le Dous, who studied experience economy and engineering, applied for the Take Off *program with their business idea, Playground Marketing. Their concept uses interactive marketing to engage the customer/user and both students made good use of the entrepreneurial environment at SVAA. Morten and Lasse have now finished their studies and are working full time on Playground Marketing, also employing a developer.*

By Stage 4, all the students are full-blown entrepreneurs, already earning a profit and focusing on developing the start-up into an established company with employees. Most students applying to the *Growth* program started as nascent entrepreneurs in the *Take Off* program. However, as Peter Kenney's example illustrates, it is possible to jump straight into the *Growth* program if the business concept is particularly innovative and high potential.

11.3.2.4 Defining characteristics of SVAA

By providing physical office space for students, SVAA is able to bring together large groups of individuals interested in the same topic (entrepreneurship) and this contributes to building and nurturing an entrepreneurial culture. Both these factors are taken into consideration when new activities are designed. Another important aspect is the visibility of staff. Through experience, their presence at SVAA often leads to several business-related – and personal – discussions.

Other important factors are:

- Entrepreneurs as active players: our entrepreneurs function as co-creators at SVAA. They assist each other, share knowledge and experience and are required to be available to answer questions and offer advice at least one to two hours per month.

- Role models: our entrepreneurs function as role models for each other, which theory states as important.
- Network: networking between start-ups and between start-ups and external partners can also lead to the establishment of new start-ups or, at the very least, provide valuable assistance in solving challenges.
- Seriousness and demands: our entrepreneurs make demands, which helps push them towards action.
- Interdisciplinary: students from all faculties at AU can join SVAA.
- Flexibility: the balance between assistance and independence is continuously challenged and adjusted at SVAA in response to feedback and results from developing start-ups.

11.4 ENTREPRENEURSHIP EDUCATION

Entrepreneurship Education (EEd) has evolved as an academic discipline. It is no longer a question of whether it can be taught but what and how it should be taught (Karlsson and Moberg, 2013; Lautenschläger and Haase, 2011; Mwasalwiba, 2010; Kuratko, 2005). AU practices EEd as a process and method that can be taught and thereby learnt. The overall approach to EEd at AU is not only to drive the creation of student start-ups but also to equip students with the skills to solve problems and adopt an entrepreneurial mindset, so that they see the world around them from a perspective of taking action and/or responding to the demand for action. This is the essence of most, if not all, entrepreneurship courses at the university. The focus is less on 'what' students learn (or create as a part of the process) but more on 'how' they learn, with an emphasis on the connection between their academic and personal skills.

In many universities or entrepreneurship environments, the criterion used to measure their effectiveness is how many start-ups are created. With its emphasis on the process, rather than on 'what' results, the impact of the AU entrepreneurship approach is more difficult to measure and/or report. However, this is where the existence of a student incubator with a flexible opt in/opt out process complements the efforts of formal education.

Because most EEd courses offered at AU are short, with credits ranging from 2 to 20 (from week-long to semester-long courses) and without the requirement at the end that the student(s) should have created a start-up, the existence of SVAA benefits students who develop a business idea during such a course and want to launch their start-up while still engaging in their studies. Several students who followed entrepreneurship education

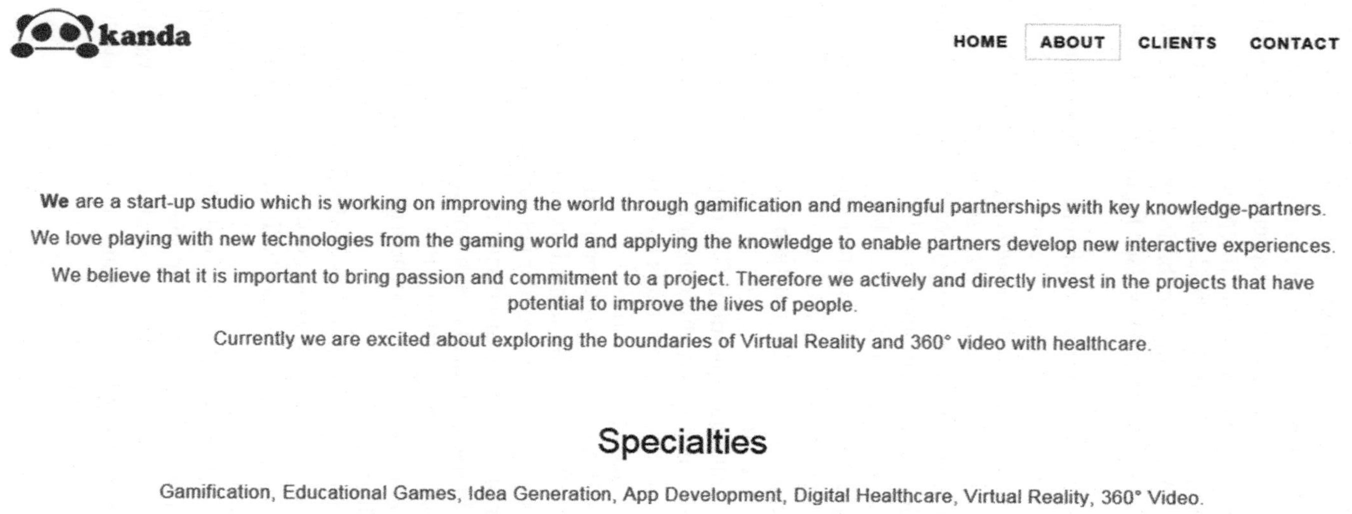

kanda

HOME ABOUT CLIENTS CONTACT

We are a start-up studio which is working on improving the world through gamification and meaningful partnerships with key knowledge-partners.

We love playing with new technologies from the gaming world and applying the knowledge to enable partners develop new interactive experiences.

We believe that it is important to bring passion and commitment to a project. Therefore we actively and directly invest in the projects that have potential to improve the lives of people.

Currently we are excited about exploring the boundaries of Virtual Reality and 360° video with healthcare.

Specialties

Gamification, Educational Games, Idea Generation, App Development, Digital Healthcare, Virtual Reality, 360° Video.

Figure 11.3 Kanda ApS start-up

at the university then contacted SVAA to take their idea forward. One such example is Kristian Andreasen from the video game start-up *Kanda ApS*. Kristian was introduced to entrepreneurship via an EEd course in the Computer Science Department. Interested in video games and the idea of using games for learning, he developed the idea for *Kanda ApS* and applied to SVAA. Following incubation in SVAA, *Kanda ApS* is now an independent and growing start-up (see Figure 11.3) with an impressive list of clients.

This example shows that the combination of exposure to EEd at university, the development of academic and personal skills, plus the existence of an incubator contributed to *Kanda ApS* being where it is today.

Referring back to Figure 11.1, we can see that EEd instilled entrepreneurial intent in Kristian but he only became a nascent entrepreneur at SVAA. This highlights the symbiotic relationship between EEd and incubation facilities. It is important to stress, however, that these are not necessarily mutually exclusive but complementary in nature because we also see examples of entrepreneurs at SVAA who entered earlier via the EAA events without exposure to EEd but then realized the need for EEd and signed up for those courses.

This raises the following question: Can entrepreneurial success for students come from either intervention alone? The short answer is yes. EEd can inspire students to create start-ups from the classroom and then go it alone (remember that start-up criteria are not a pre-requisite of EEd at AU). Similarly, SVAA can and has supported students towards successful entrepreneurship. But the complementary nature of both interventions means that AU is able to cast a broader net across the university and reach a higher number of students. Considering the size of the university, this is an achievement in itself. Students are more likely to be mentally prepared for the challenges entrepreneurship will throw at them because they are assisted in becoming stronger entrepreneurial individuals, as the experience of *Kanda ApS* shows us.

This complementarity not only ensured a broad reach but also contributed to the critical mass SVAA needed to create the right growth and learning environment for students. EEd also provides cutting-edge knowledge from entrepreneurship research that SVAA can apply in practice. Collaboration between educators and SVAA practitioners is another positive offshoot: educators learn from the 'realization' activities students engage in while practitioners learn from and apply relevant theories and best practices in more-or-less real time. This collaboration has been a tangible boost for experiential learning.

11.5 OVERALL IMPACT AND FUTURE PERSPECTIVES

The concrete outcomes of SVAA (for the period April 2011 to February 2014) are as follows:

- Participants in open entry programs:
  - *Sand Box*: 198 students (eight courses in total)
- *Entrepreneurs' Workshop*: 72 students (two courses in total)
- Applicants to the *Take Off* program *Explore*: 296 students
- Participants in *Take Off* (*Explore*, *Incubator*, and *Growth*): 208 concepts/start-ups (329 students)
- Number of start-ups (defined by Central Virksomhed (company) Registration number (CVR registrations)): 124
- For start-ups participating in the *Growth* program:
  - Annual turnover (2013): 460,400 euros (based on the 14 start-ups presently in *Growth*)
  - Number of jobs created: 25 partners/owners, 11 full-time employ ees, 41 part-time employees = 77 in total (based on 21 start-ups in *Growth* during 2013)
- Applicants for *Corporate Take Off*: 99 students and 22 established companies
  - Matches made: 20 (58 students and 16 companies)
- Entrepreneurship activity level of former SVAA participants (response rate: 64.4%)
  - Percentage still an entrepreneur: 53.1%
  - Percentage with entrepreneurial intent (no start-up as yet): 35.4%

11.5.1 Impact

While the above numbers speak for themselves, what is the concrete impact of SVAA on two of its most important stakeholders: the students and the university?

Students: The focus on individual development and its interplay with program design has first and foremost had an impact on the appeal of entrepreneurship for university students. This appeal is growing exponentially (measured by the increasing number of applications to SVAA and the positive feedback from students). We do not take full credit for this increased interest. Favorable 'market conditions', such as positive media coverage of entrepreneurship and easier communication and internet tools, have also contributed to this exponential growth. However, we can safely say that our initial work in establishing and embedding an entrepreneurial culture at AU coupled with a flexible, broad-based program design with multiple entry points enabled SVAA to be 'market-ready'. Now when students actively look for entrepreneurial opportunities they are more aware of our presence. The impact of the numbers above is even more relevant if one correlates the data with the entrepreneurial journey flowchart (Figure 11.1).

University: The same data also illustrate the impact on AU. But more importantly, the positive movement of entrepreneurship into the mainstream has translated into increased support from the university leadership for entrepreneurial activities and learning. This has resulted in more entrepreneurship courses alongside complementary courses (traditional courses with an entrepreneurial orientation), which has broadened the uptake of entrepreneurship and led to an increased number of applications to SVAA. The effects of increased entrepreneurial education at the university are still being measured, but the experience of AU clearly supports the claim that 'Entrepreneurship has moved from whether it can be taught to how it should be taught' (Mwasalwiba, 2010). Furthermore, the university is taking a more active interest and stake in building an entrepreneurial ecosystem both inside and outside the university. The best example of this is SVAA's funding, which has now moved from project financing (DEU) to basic funding by the university. This effectively frees SVAA to focus on its core activities (generating value for students and the university) as opposed to operating within the constraints of a project funding infrastructure.

Aside from these direct impacts on our primary stakeholders, other impacts on both students and their learning are more difficult to measure. However, the premise of an increased entrepreneurial mindset and the spillover effects on the development of innovation and an entrepreneurial ecosystem in the region cannot be discounted. Similarly, intrapreneurship in small and medium-sized enterprises (SMEs) can have a positive impact on regional growth and innovation. We already see evidence of this in our collaboration with a parallel activity at AU called SpinIN. This project, financed by the Danish Industry Foundation, matches SVAA entrepreneurs and/or start-ups with established SMEs and is growing in popularity.

11.5.2 The Future of SVAA

As explained above, SVAA is now fully integrated in AU and the university funds almost all of its activities. All AU students can apply to SVAA but, because of growing interest, we have now implemented an application process. This means we are not able to support all applicants. However, in the future we hope to expand our services to meet the growing demand from students.

A wide range of activities and workshops are open to anyone with an interest in entrepreneurship. In the future, SVAA hopes to contribute to the wider entrepreneurial ecosystem in Denmark, helping to create a larger network that will benefit AU students, researchers and the university as a whole.

Another priority for SVAA is continuing to actively support the university's strategy in the following areas:

(1) positioning AU as an attractive place to study for the most talented students in Denmark and abroad;
(2) equipping AU students with strong academic, innovative and entrepreneurial skills that meet the demands of business and society; and
(3) contributing to societal growth and development.

SVAA now has more than six years of experience. Each year we collect qualitative and quantitative data in order to document and measure our impact and to better understand the effects of a student incubator experience as a part of a student's academic career. As entrepreneurs ourselves, we are continually developing new methods, exploring new ways of approaching students, and new ways of collaborating with the local entrepreneurial ecosystem, etc., to find the best possible business model for the university and its student entrepreneurs.

One thing that is unlikely to change is our mission statement: '*We develop the individual to qualify the individual to develop his/her company*'. This remains of key importance due to the embedded value this provides for all stakeholders and is reflective of our approach towards student incubators. The student incubator is a learning environment where students are given the opportunity to become entrepreneurs with the option to gain, fail or leave (with no-strings-attached). The start-up and entrepreneurial environment at SVAA provides the framework around which these skills are learned. The start-up is therefore a tool – a way for students to learn how to put their academic knowledge and personal strengths into practice. The core task of SVAA will always be to support students in their development as competent and talented individuals with high academic skills, who are able to make a difference and add value to society – as intrapreneurs or entrepreneurs – and to their lives in general.

NOTE

1. The latter, *Entrepreneur Appetizer*, is currently being developed based on a former program, *Sand Box*.

REFERENCES

Bailetti, T. and J. Doerr (2011), 'Fostering student entrepreneurship and university spinoff companies', *Technology Innovation Management Review*, **2011** (10), 7–12.

Blank, S. and B. Dorf (2012), *The Startup Owner's Manual: The Step-by-Step Guide for Building a Great Company*, Vol. I (Pescadero, CA: K&S Ranch, Inc).

Davidsson, P. (1995), 'Determinants of entrepreneurial intentions', *RENT IX Workshop in Entrepreneurship Research*, Piacenza, Italy.

Davidsson, P. (2006), 'Developments in the study of nascent entrepreneurs', *Foundations and Trends in Entrepreneurship*, **2** (1), 1–76.

Heinonen, J. and U. Hytti (2010), 'Back to basics: the role of teaching in developing the entrepreneurial university', *International Journal of Entrepreneurship and Innovation*, **11** (4), 283–92.

Jeraj, M. and B. Antoncic (2013), 'A conceptualization of entrepreneurial curiosity and construct development: A multi-country empirical validation', *Creative Resources Journal*, **25** (4), 426–35.

Karlsson, T. and K. Moberg (2013), 'Improving perceived entrepreneurial abilities through education: Exploratory testing of an entrepreneurial self-efficacy scale in a pre-post setting', *International Journal of Management Education*, **11** (1), 1–11.

Kirketerp, A. (2011), 'Foretagsomhedsdidaktik – Skubmetoden', in Anne Kirketerp and Linda Greve (eds), *Entreprenørskabsundervisning* (Aarhus: Aarhus Universitetsforlag), 93–125.

Kuratko, D.F. (2005), 'The emergence of entrepreneurship education: Development, trends, and challenges', *Entrepreneurship Theory and Practice*, **29** (5), 577–97.

Lautenschläger, A. and H. Haase (2011), 'The myth of entrepreneurship education: seven arguments against teaching business creation at universities', *Journal of Entrepreneurship Education*, **14**, 147–61.

Michelacci, C. (2003), 'Low returns in R&D due to the lack of entrepreneurial skills', *The Economic Journal*, **113** (484), 207–25.

Mwasalwiba, E.S. (2010), 'Entrepreneurship education: a review of its objectives, teaching methods, and impact indicators', *Education Training*, **52** (1), 20–47.

Osterwalder, A. and Y. Pigneur (2010), *Business Model Generation: A Handbook for Visionaries, Game Changers, and Challengers* (Hoboken, New Jersey: John Wiley & Sons).

Sarasvathy, S.D. (2009), *Effectuation: Elements of Entrepreneurial Expertise* (Cheltenham, UK and Northampton, MA: Edward Elgar Publishing).

PART 5

International perspective on academic spin-offs and technology transfer

12. What Europe still has to learn from the US in academic innovation

Hervé Lebret

Can we learn anything from ongoing high-tech success stories coming from the US in the last half-century? Europe and the rest of the world have tried to understand and copy technology clusters such as Silicon Valley and the Boston Area without much success. It is now generally acknowledged that it is not possible to copy and paste the American model on the Old Continent, but that does not mean that efforts should not be renewed. New lessons might be learnt from these failures. First, technology innovation is a culture; it is not a process that can be planned. Secondly, high-tech entrepreneurship requires a combination of human and financial resources, which are very different from those needed in traditional industries. If we can agree upon these lessons, Europe might be in a position to fight back.

12.1 AMERICAN LEADERSHIP IN HIGH-TECH INNOVATION IS NOT A MYTH

There is a sound exercise that anyone curious about high-tech innovation should do once a year: which of today's US technology giants were small start-ups less than 50 years ago? And then, who are their European counterparts? Whatever the metrics (sales, profits, employment, market value), the top ten American success stories would probably be the same and Table 12.1 is an illustration of these in the information technology sector. Obvious names spring to mind such as Apple, Microsoft, Oracle and Intel, which are all around 40 years old or more, or Google, Amazon, Facebook and Twitter, which are less than 20 years old. Dell or Sun would have been included if the list had been made in 2010, whereas Facebook and Twitter would be newcomers.

Some of these numbers speak for themselves: the value creation is huge and so is the number of employees even if Facebook, Twitter or even Tesla Motors have more modest headcounts. Hewlett-Packard (HP) is mentioned not only as a reference but also to show that not all technology

Table 12.1 US giants in information technology

Company	Founded	IPO	Market value	Employees
Apple	1976	1980	$520B	80,000
Google	1998	2004	$367B	50,000
Microsoft	1975	1986	$327B	99,000
Oracle	1977	1986	$185B	120,000
Facebook	2004	2012	$150B	7,000
Amazon	1994	1997	$138B	117,000
Qualcomm	1985	1991	$135B	31,000
Intel	1968	1971	$130B	107,000
Cisco	1984	1990	$124B	75,000
eBay	1995	1998	$66B	32,000
Yahoo	1994	1996	$34B	12,000
Tesla Motors	2003	2010	$24B	6,000
Twitter	2006	2013	$18B	3,000
Average	**1989**	**1994**	**$171B**	**57,000**
HP	1939	1957	$62B	317,000

Source: *Yahoo Finance*, 21 May 2014

companies are young corporations. A less obvious output of this data is that it only took on average five years from their foundation for these companies to enjoy an initial public offering (IPO). A similar exercise could have been done in the life sciences (biotechnology and medical devices) with similar results[1] (even if with smaller numbers). Now, the same exercise needs to be done for Europe and the reader should be aware that it will be much more challenging. The US probably has tens of companies with billion-dollar value whereas Europe has a very limited number of 'young' companies selling high-tech products with such value. SAP is one exception and can be compared to the American giants, but even the second company on the European list would have been at the bottom of the US list. Again, the numbers are eloquent and the number of years from incorporation to IPO is about twice as long as in the US. Nokia is added for reference just as HP was for the US, but Nokia is a very old company even if it was recently reborn with the advent of mobile telecommunications.

There are other lessons to be learnt from comparing these two groups of 13 companies. Ten out of the 13 US companies were initially funded with venture capital (VC) before they went public. Only five in the European group were VC-backed. The age of founders is another element of interest. With the exception of Intel and Qualcomm, whose founders had created previous companies more than ten years earlier, the average age[2]

Table 12.2 European leaders in information technology

Company	Founded	IPO	Market value	Employees
SAP	1972	1988	$88B	66,000
ASM Litho.	1984	1995	$35B	10,000
ARM Holding	1990	1998	$20B	3,000
Dassault Syst.	1981	1996	$15B	11,000
Gemalto	1988	2000	$9B	12,000
King Digital	2002	2014	$5B	800
Logitech	1981	1990	$2B	9,000
Criteo	2005	2013	$2B	800
Betfair	2000	2010	$1.7B	1,600
CSR	1998	2004	$1.5B	2,000
Swissquote	1997	2000	$0.7B	500
Soitec	1992	1999	$0.6B	1,200
F-secure	1988	1999	$0.4B	900
Average	**1990**	**2000**	**$14B**	**9,000**
Nokia	1965 (1966)	1915 (1994)	$27B	55,000

Source: *Yahoo Finance*, 21 May 2014

of American founders is 27. The average age of European founders is 33. Could it be that fast growth and youth are somehow related?

12.2 US ACADEMIC INSTITUTIONS ALSO LEAD IN ENTREPRENEURSHIP

US universities are not much different. They lead the world in academic ranking and most recently Harvard, Berkeley, Stanford and MIT occupied the top four places.[3] Zhang (2003, 2009) showed exactly the same metrics for entrepreneurship. Table 12.3 is an illustration of university-led entrepreneurial creation. It shows that MIT and Stanford, with their more technical orientation, created the most VC-backed companies in the 1990s. More specific studies about MIT (Roberts and Eesley, 2011) and Stanford (Lebret, 2010; Eesley and Miller, 2012) dig into the value creation by these companies, the resources available (financial, human, ecosystem support) for their growth as well as the demographics of the founders. These studies mention millions of jobs created, billions of dollars in funding and trillions of dollars in revenues. The support systems are also explored with an emphasis on education and exposure to entrepreneurship as a first step,

Table 12.3 Origin of founders in the two main technology clusters

VC-backed Founders from Leading Institutions in Silicon Valley and in the Boston Area (1992–2001)			
Origin of Founders	Silicon Valley (1)		Boston Area (2)
Leading Companies			
Apple	94	Data General	13
Cisco	41	DEC	52
HP	117	EMC	9
Intel	76	Lotus	29
Oracle	73	Prime	5
SGI	50	Raytheon	7
Sun	101	Wang	11
IBM	82	IBM	23
Leading Universities			
Stanford	71	MIT	74
UC Berkeley	20	Harvard	32

Note:
(1) Founder sample size: 2,492
(2) Founder sample size: 1,157

Source: Zhang, 2003

high-quality networks for entrepreneurs as a second step, with high (even if competitive) availability of capital as a third step.

The analysis for European universities is even worse than for corporations. Only three universities (Cambridge, Oxford and ETH Zürich) belong in the top 20. In the field of entrepreneurship, the situation is similar, with much fewer success stories. The analysis of ETH spin-offs (Oskarsson and Schläpfer, 2008) shows an interesting focus on firm survival more than on value creation, which is common to most European academic institutions. A strange consequence is that US academic start-ups have a 50% survival rate after five years whereas their European counterparts reach 90%, which is very unusual in traditional economic activities. There is no focus on fast growth of emerging technology leaders.

12.3 INNOVATION IS A CULTURE

Despite the apparent simplicity of this initial description, it would be a big mistake to consider that innovation is a (simple) process. In an analysis of Silicon Valley as an entrepreneurial ecosystem, Evans and Bahrami

(2000) list the necessary ingredients of technology clusters: universities and research centers of the highest caliber; an industry of VC (i.e. financial institutions and private investors); experienced professionals in high-tech; service providers (such as lawyers, head hunters, public relations and marketing specialists, auditors, etc.); and last but not least, *an intangible yet critical component*: a pioneering spirit that encourages an entrepreneurial culture. If the first four ingredients are easy to implement, the last one implies a much more challenging change of culture.

Table 12.3, which already indicates the importance of academic institutions, delivers another important message. Entrepreneurs do not only come from academic institutions. They also spin out of technology firms. Zhang compares Silicon Valley and the Boston Area. There is not much difference academic-wise, but the situation is strikingly different for firms created by industry people. The case of IBM on the West and East Coast is a good example. In her now classic *Regional Advantage: Culture and Competition in Silicon Valley and Route 128*, AnnaLee Saxenian (1994) analyzed these differences and showed that the main differences were cultural. A technology giant such as Digital Equipment (DEC) had been built on a traditional corporate culture where confidentiality was critical and leaving your employer was close to treason. In comparison, the Eight Traitors who left Shockley Labs to found Fairchild Semiconductor in 1957 are celebrated as the fathers of Silicon Valley. Even if only as an anecdote, it is worth mentioning the now defunct Wagon Wheel Bar: 'During the 1970s and 1980s, many of the top engineers from Fairchild, National and other companies would meet there to drink and talk about the problems they faced in manufacturing and selling semiconductors. It was an important meeting place where even the fiercest competitors gathered and exchanged ideas'.[4] This is what open innovation is really about!

Paul Graham, a Silicon Valley entrepreneur and founder of the Ycombinator accelerator, offers the same if even more provocative analysis: 'I read occasionally about attempts to set up "technology parks" in other places, as if the active ingredient of Silicon Valley were the office space. An article about Sophia Antipolis bragged that companies there included Cisco, Compaq, IBM, NCR, and Nortel. Don't the French realize these aren't startups?' and he claimed in the same article: 'Few startups happen in Miami, for example, because although it's full of rich people, it has few nerds. It's not the kind of place nerds like. Whereas Pittsburgh has the opposite problem: plenty of nerds, but no rich people'.[5] Silicon Valley is simply this combination of nerds and rich people, of entrepreneurs and investors.

12.4 BOULEVARD OF BROKEN DREAMS

Josh Lerner is one of the top experts in high-tech innovation and entrepreneurship. He analysed in his *Boulevard of Broken Dreams* (2009): 'the common flaws undermining far too many programs – poor design, a lack of understanding for the entrepreneurial process, and implementation problems'. Lerner explains why governments cannot dictate how venture markets evolve, and why they must balance their positions as catalysts with an awareness of their limited ability to stimulate the entrepreneurial sector. As governments worldwide seek to spur economic growth in ever-more aggressive ways, he offers an important caution and proposes a careful approach to government support of entrepreneurial activities, so that the mistakes of earlier efforts are not repeated.

It is not sufficient to expose, teach, offer office space and incubate with financial resources and industry expertise to build efficient technology clusters. Entrepreneurship is about great determination in the face of uncertainty and a high risk of failure. 'Launching a start-up is not a rational act. Success only comes from those who are foolish enough to think unreasonably. Entrepreneurs need to stretch themselves beyond convention and constraint to reach something extraordinary',[6] said Vinod Khosla, co-founder of Sun Microsystems and then a venture capitalist with Kleiner Perkins.

The main mistake or underestimation of the difficulty in building entrepreneurial ecosystems has been to set in place the basic tools considered necessary in an often top-down approach with the idea that entrepreneurship will somehow develop naturally. Most European universities have created technology transfer offices and entrepreneurship courses; they offer office or incubation space, coaching programs and sometimes they even have venture funds to invest directly in their spin-offs. These are not wrong approaches, as Lerner correctly showed, but their success is totally dependent upon their implementation and the leadership of the professionals managing these tools; and, more importantly, success is dependent upon the ambition of their entrepreneurs. This ambition is often misplaced and leads to the 90% survival rate of European start-ups, which can only be explained by a fear of failing instead of a risk-taking attitude focusing on high value creation.

12.5 EUROPE HAS NOT UNDERSTOOD HIGH-TECH ENTREPRENEURSHIP FOR DECADES

Some of the European leaders in technology mentioned in Table 12.2 were not typical start-ups. ASML was a spin-off from Philips, Dassault Systems

Table 12.4 Comparison of the traditional VC models in the US and in Europe

	US Model	Old European Model
People	Entrepreneurs (founders and builders)	Consultants and accountants
Stage	Creation (seed)	Early (A round) but not seed
Provide	'Value added'	'Just money'
Style	Hands on (active)	Hands-off (more passive)
Talent	Sales & marketing	Scientists
Objective	Create very large companies	Create medium size companies
Philosophy	Maximize upside	Minimize downside
Returns	Target a small number of big winners (home run investing)	Believe returns can be earned across the portfolio

was a spin-off from Dassault Aviation, and ARM was a joint venture of Apple. Most of these companies were closely linked to established corporations. If Zhang and Saxenian are right about the necessary open corporate culture, they provide an explanation for Europe's failure. Similarly, European VC was often associated with traditional banks and consulting firms, whereas the pioneers of the US VC world were former entrepreneurs from the first Silicon Valley start-ups. In 2006, Tim Cruttenden made an interesting comparison at the International Venture Capital Conference (see Table 12.4).

Lessons were learnt, thanks in particular to successful European entrepreneurs. An example is given by Daniel Borel, co-founder of Logitech:

> The only answer I may provide is the cultural difference between the US and Switzerland. When we founded Logitech, as Swiss entrepreneurs, we had to quickly enter the international scene. The technology was Swiss but the US, and later the world, defined our market, whereas production quickly moved to Asia. I don't want to sound too definitive because things change and many good things happen in Switzerland. But I do feel that in the US, people are more open. When you receive funds from venture capitalists, you automatically accept an external shareholder who will help you manage your company and who may even fire you. In Switzerland, this vision is less well accepted: One prefers to fully control a small pie than only control 10% of a big pie, and this may be a limiting factor.[7]

After decades of misunderstanding, during which entrepreneurs and investors thought that building a European company was a process of imitating the traditional industries they were close to, these same entrepreneurs who were exposed to the American model have helped to change the situation. Daniel Borel would certainly not have enjoyed the success of Logitech without years spent at Stanford University and in Silicon Valley. Bernard Liautaud, the co-founder of Business Objects, which would have belonged to the list of European leaders if it had not been acquired by SAP for more than US$6 billion, offers a similar analysis: 'we immediately launched the company by copying the entrepreneurial models of Silicon Valley'. But he adds: 'when BO went public in 1994, we thought we would be followed by other European actors . . . I have been disappointed by the fact that there were finally few others'.[8]

The last decade has seen a multiplication in European high-growth companies. Skype is probably the best-known success story. In a recent article for the *Financial Times* (FT), Skype's co-founder, Niklas Zennström, stated: 'Although it is possible to build a global business much quicker than ever before, it is still not easy. The skills needed to develop a great product are not the same as needed to build and run a global business'.[9] Risto Siilasmaa, F-secure co-founder and today chairman of Nokia, also understands these new values: 'Entrepreneurship should be cherished, because it will be critical for the future of the world. It is not a profession; it is a state of mind'.[10]

12.6 EUROPE WILL NEVER HAVE A SILICON VALLEY AND THAT'S GOOD NEWS

In the FT article cited above, Niklas Zennström wrote:

> Silicon Valley is being challenged by cities from London to Berlin and Beijing . . . In fact, 10 years ago when we were building Skype from Sweden, one potential investor told me that he would invest on condition I moved to Silicon Valley. I did not move and he did not invest. . . . Today, the simple truth is that great companies can come from anywhere. In a way, we have the wrong obsession with geography. In technology, as in life, it is not where you come from – it is where you are going that counts.

Zennström co-founded Atomico, a venture capital firm. Liautaud is a partner with Balderton, another London-based VC firm. Daniel Borel is a powerful business angel in Switzerland. Dominique Vidal, co-founder of Kelkoo, a start-up acquired by Yahoo! in 2005 for US$500 million became a partner with Index Ventures, which has invested in Criteo, Supercell and previously in Skype. The analysis of Cruttenden in Table 12.4 needs to be updated with a new European model similar to the US one, where Index Ventures, Atomico, Balderton and others funded Skype, Betfair, Rovio, Supercell, Kind Digital or Criteo, the most recent European success stories with a typical American model.

The remaining challenge is that these success stories are dispersed across Europe and no real cluster has or will ever emerge. Paris, London and Berlin are fighting for leadership as much as other smaller cities. Efforts might be diluted and wasted. But the good news is that a new generation of extremely ambitious entrepreneurs and investors has emerged. Tom Perkins (from legendary VC fund Kleiner Perkins) is absolutely right in his analysis:[11]

> The difference is in psychology: everybody in Silicon Valley knows somebody that is doing very well in high-tech small companies, start-ups; so they say to themselves: 'I am smarter than Joe. If he could make millions, I can make a billion'. So they do and they think they will succeed and by thinking they can succeed, they have a good shot at succeeding. That psychology does not exist so much elsewhere.

No other geographic area has the same concentration of talents who became role models for the next generation. It is a little known fact that Robert Noyce, founder of Intel, was a mentor, not to say a second father, to Steve Jobs. However, at European level, important role models are emerging for the next generation of entrepreneurs. These successful entrepreneurs are becoming mentors and investors, advising young entrepreneurs on how to become ambitious global leaders.

12.7 INCUBATORS, ACCELERATORS, CO-WORKING SPACES: A NEED FOR NEW TOOLS?

Culture, talent and money – that is what innovation is all about. It is what Paul Graham simply described as the critical and necessary ingredients of high-tech entrepreneurship. So why did the same Paul Graham create StartupSchool (a conference with famous speakers for young entrepreneurs and developers) and Ycombinator (an accelerator in the software and internet fields) in 2005? Whereas 15 years ago incubators were focusing on helping entrepreneurs in the early development of their start-ups, co-working spaces today add the concept of connecting people with different expertise and projects on top of which accelerators offer intense training programs for these young developers and entrepreneurs. Quick feedback from experts is given to the teams who can immediately adapt and improve their business projects. Ycombinator has been copied all over the world and in fields outside of software. The basic concept still remains the gathering of talents in a very competitive environment, where collaboration is encouraged. The camp-like nature encourages intensity, fast reactions and ambition; these are the necessary ingredients for giving the start-ups an opportunity to grow. Seedcamp and Startupbootcamp are the early European leaders and their names are an indication of the same culture. These camps are the new Wagon Wheel Bars.

Even if many such accelerators have developed in all major American cities, it is interesting to note that Paul Graham closed his activity in Boston (where it was initially launched): 'Boston just doesn't have the startup culture that the Valley does', Graham writes, 'It has more startup culture than anywhere else, but the gap between number 1 and number 2 is huge'. In Europe, Seedcamp moves location for each session and takes place in such diverse places as London, Athens, Tallinn, Berlin, Zagreb, Barcelona, Budapest, Lisbon and Paris. Both Seedcamp and Ycombinator are sponsored by business angels and venture capitalists. Both had early success stories, but out of the 300+ Ycombinator companies and the 70+ Seedcamp companies, only a few have become famous, such as Airbnb and Dropbox. It is probably too early to say whether any giants will emerge from this new concept.

The claimed advantages of these camps are strong and easy connections for developers with peers and industry experts. But despite the argument that software companies can be launched in short periods of time with very little capital, the truth is that these programs are very efficient filters used by venture capitalists to select promising young entrepreneurs and projects. The subsequent growth of these start-ups follows the traditional Silicon Valley VC trajectory, even in Europe, which has now clearly understood and adopted the Anglo-Saxon model. There's not really anything new under the entrepreneurial sun except the consolidation of a tried and tested model! Connecting talent with money. But for Europe, adopting the accelerator model is recognition that it is the best way to counter the isolation of young entrepreneurs and innovators. Instead of fighting for centrality, they decentralize themselves all over Europe.

12.8 FAVORING EXCHANGES THROUGH DECENTRALIZED ACCELERATION

What Europe has succeeded in doing with the Erasmus project in university education could be leveraged in entrepreneurship via accelerators. Universities are still critical once students have left, providing landing zones for mobile entrepreneurs, which is all the more important if Europe cannot have its own Silicon Valley. Without accelerators, Europe's entrepreneurs risk becoming even more isolated. Europe needs to support high-energy mobility among entrepreneurs between European hotspots, just as it does so successfully for its students.

Aalto University recently launched the Startup Sauna program 'with a mission to build a functioning startup ecosystem in Northern Europe'. Many European universities are exploring the opportunity of creating joint accelerators and have learnt the lesson from Silicon Valley that academic entrepreneurship cannot be developed outside of the broader entrepreneurial ecosystem. Aalto also included Silicon Valley as well as VC connections in its Startup Sauna program. Even if the ambition of the EU's Lisbon agenda to make Europe 'the most competitive knowledge-based economy' by 2010 has been a total failure, in part because of all the misunderstandings explained here, the Horizon 2020 program is an opportunity to support a new ambition through Europe's leading universities and their entrepreneurs.

The European Startup Manifesto[12] initiative, launched in 2013 by the founders of Skype, Spotify, Rovio and Seedcamp, is the best expression of this ambition (p. 1):

> The days of relying on large businesses or the government for job creation are over. Many of the millions of jobs lost over the past five years will never return in their old form. Entrepreneurship, which has been the engine for growth in the United States, has not been cultivated in an effective or systematic way in Europe. To create more businesses and more startups requires more than a change in policy. It requires a change in mentality.

The Manifesto's 22 recommendations are unsurprisingly gathered in five groups including education, access to talent, access to capital and thought leadership (the last domain being about a common European policy on data protection and privacy). Europe's leading universities can be game changers here – catalysts – by agreeing on what is important and investing in it.

12.9 THE ENTREPRENEURIAL STATE

It would be a fatal mistake to believe that these necessary paradigm changes could revolutionize the fundamental missions of governments and universities. In a formidable analysis of the importance of the public sector in entrepreneurship and innovation, Mariana Mazzucato (2013) showed that the US became the world leader in technology thanks to fantastic public spending in education and research, which went as far as defining new innovations that private companies could successfully commercialize. Even if they are often private entities, American universities are mostly sponsored by government agencies for their research. Many breakthrough innovations were created in academic laboratories with public funding, for example, the internet by DARPA and many new drugs by NIH, before being licensed to private companies.

Though it can be argued that Mazzucato discounts the critical role of individual entrepreneurs far too much, she soundly reminds us that the US was entrepreneurial in its adventurous risk-taking attitude towards exploring new technological innovations and was not rewarded for its investments in comparison to private entrepreneurs and investors. Not everyone will support her claim that the public sector should now take the leadership that private investors have abandoned in the early stages of technology development, but these fundamental roles and missions of the state are nicely restated and should not be abandoned, certainly not in Europe.

12.10 THE IMPORTANCE OF AGE IN DISRUPTIVE AND INCREMENTAL INNOVATION

Accelerators are the extreme expression of the importance of age. The average age of entrepreneurs at Ycombinator is 26 and is likely to be similar in other accelerators. However, a debate has been raging about the importance of youth, particularly after a study by the Kaufmann foundation[13] showed that the average age of entrepreneurs was above 40. Experience matters too and explains why many entrepreneurs are middle-aged. The reason why youth is important is different. As Drew Hanson[14] suggests:

> The conceptual innovator generally conceives his ideas in their entirety before beginning production. They boldly break away from accepted standards, but after their early breakthroughs, tend to fade away, or at least their later work never attains the height of their early success. Experimental innovators create in a different manner. It takes time for them to hone their craft. They return to familiar ideas, trying to perfect them. While conceptual innovators are more deductive in their approach, experimental innovators are inductive and use their observational powers to infer and concoct hypothesize. They use trial-and-error, and it's this process, not the final product, that fascinates them the most. They devote their life to learning, seeking answers to their unsolved questions.

Hanson could have used the well-known concepts of disruptive vs. incremental innovation. This analysis might explain why Europe has fewer disruptive innovations and fewer young entrepreneurs.

Hanson also mentions Noyce as a middle-aged entrepreneur (although he forgets that Noyce founded Fairchild when he was 30). Here is a quote by Don Valentine (2002), one of the best known Silicon Valley venture capitalists, about a young Steve Jobs and his 'older mentor' Noyce:

> There are only two true visionaries in the history of Silicon Valley. Jobs and Noyce. Their vision was to build great companies . . . Steve was twenty, undegreed, some people said unwashed, and he looked like Ho Chi Min. But he was a bright person then, and is a brighter man now . . . Phenomenal achievement done by somebody in his very early twenties . . . Bob was one of those people who could maintain perspective because he was inordinately bright. Steve could not. He was very, very passionate, highly competitive.

And Valentine adds as a common characteristic: 'Founders are genetically impossible by choice'.

Similar to artists and scientists, breakthrough innovators seem to be young, and prodigies often express themselves best in their 20s. More experienced creators also exist but might produce more incremental work. When experience is involved, the need for role models, mentors, and good ecosystems is less obvious. But young people with no experience and a lot of talent need to be nurtured and encouraged, surrounded by friendly and valid advice. Accelerators claim to provide this much needed access to experience and capital. In 2006, an analysis by the British Library House, *Beyond the Chasm*, identified that only 2% of British VC-backed start-ups were managed by entrepreneurs under the age of 30.

It is not so much that youth is a necessary ingredient of success but that youth can be an asset. Young people should be trusted and never replaced by more experienced people. They should be surrounded and advised by experienced people in order to increase their chances of success. Mike Markkula, a former Intel executive, provided such support to Steve Jobs and Steve Wozniak when they had no business experience. Eric Schmidt became Google's CEO a few years after its foundation but never replaced Sergey Brin or Larry Page.

12.11 CELEBRATING ENTREPRENEURS

As Risto Siilasmaa said, 'Entrepreneurship should be cherished'. Robert Noyce, Steve Jobs' mentor, went even further: 'Look around at who the heroes are. They aren't lawyers, nor are they even so much the financiers. They're the guys who start companies'. A recurring mistake of aspiring technology clusters is to put too much emphasis on the richness of their support tools. In an ideal world, the support mechanisms should have the ambition of disappearing in the long term and, at the very least, becoming invisible. When Daniel Borel launched Logitech, there was not much support around. Venture capital was in its early days when Intel and Apple were founded. But the ambition of the entrepreneurs was limitless. The skeptical reader should watch the great documentaries *Something*

Ventured,[15] which describes the early days of Silicon Valley, and *Startup Kids*,[16] which is about the new generation of Web 2.0 entrepreneurs. They are perfect illustrations of how the right combination of culture, talent and money enables successful innovations despite the high rate of failure and huge uncertainties on technology markets. A friendly infrastructure should reduce the fear of trying. Does a child successfully ride a bicycle on its first attempt? Again, youth are creative because they have yet to experience failure. It is another reason why innovation is bottom-up, and cannot or should not be top-down. T.J. Rodgers, a famous Silicon Valley entrepreneur, said 'failure is a prerequisite to success'. A Chinese student of mine seconded this with 'Shi Bai Nai Cheng Gong Zhi Mu' ('failure is the mother of success'). We need role models because we must allow our younger generation to dream. We must allow them to live their passions so that they can show us the way forward. These last sentences are less policy recommendations than slogans in favor of entrepreneurship!

NOTES

1. http://www.startup-book.com/2010/06/14/europe-vs-usa-growth-in-it-and-biotech/ (accessed 5 May 2016).
2. Author's own data and Lebret (2014).
3. http://www.shanghairanking.com (accessed 5 May 2016).
4. http://www.sfgate.com/business/article/TECH-CHRONICLES-A-daily-dose-of-postings-from-2592407.php (accessed 5 May 2016).
5. http://www.paulgraham.com/siliconvalley.html (accessed 5 May 2016).
6. http://ecorner.stanford.edu/authorMaterialInfo.html?mid=25 (accessed 5 May 2016).
7. Translated from http://www.largeur.com/?p=3016 (accessed 5 May 2016).
8. http://www.oezratty.net/wordpress/2006/rencontre-avec-bernard-liautaud (accessed 5 May 2016).
9. http://www.ft.com/intl/cms/s/0/156569c4-6c06-11e3-85b1-00144feabdc0.html#axzz2rsUrDN1Z (accessed 5 May 2016).
10. http://www.startup-book.com/2012/09/17/entrepreneurship-should-be-cherished-the-nokia-chairman-says/ (accessed 5 May 2016).
11. http://www.venturevoice.com/2007/12/vv_show_47_tom_perkins_of_klei.html (accessed 16 May 2016).
12. http://startupmanifesto.eu/ (accessed 5 May 2016).
13. http://www.kauffman.org/what-we-do/research/2010/05/the-anatomy-of-an-entrepreneur (accessed 5 May 2016).
14. http://www.forbes.com/sites/drewhansen/2012/12/04/why-arguing-about-an-entrepreneurs-age-misses-the-point/ (accessed 5 May 2016).
15. http://www.somethingventuredthemovie.com/ (accessed 5 May 2016).
16. http://thestartupkids.com/ (accessed 5 May 2016).

REFERENCES

Cruttenden, Tim (2006), 'European Venture Capital: What's Driving Investment Trends?', International Venture Capital Conference, Melbourne.

Eesley, C.E. and W. Miller (2012), 'Stanford University's Economic Impact via Innovation and Entrepreneurship', http://news.stanford.edu/news/2012/october/innovation-economic-impact-102412.html (accessed 6 November 2014).

Evans, S. and H. Bahrami (2000), 'A Flexible Recycling', in M. Kenney (ed.), *Understanding Silicon Valley, the Anatomy of an Entrepreneurial Region* (Stanford University Press) 165–89.

Lebret, H. (2010), 'Stanford University and High-Tech Entrepreneurship: An Empirical Study', http://ssrn.com/abstract=1983858 (accessed 5 May 2016).

Lebret, H. (2014), 'Age and Experience of High-tech Entrepreneurs', http://papers.ssrn.com/abstract=2416888 (accessed 5 May 2016).

Lerner, J. (2009), *Boulevard of Broken Dreams: Why Public Efforts to Boost Entrepreneurship and Venture Capital Have Failed – and What to Do About It* (Princeton/Oxford: Princeton University Press).

Library House (2006), *Beyond the Chasm: The Venture-Backed Report – UK – 2006* (Cambridge: Library House).

Mazzucato, M. (2013), *The Entrepreneurial State – Debunking Public vs. Private Sector Myths* (New York/London: Anthem Press).

Oskarsson, I. and A. Schläpfer (2008), 'The Performance of Spin-off Companies at the Swiss Federal Institute of Technology Zurich', ETH-Transfer.

Roberts, E.B. and C.E. Eesley (2011), 'Entrepreneurial Impact: The Role of MIT – An Updated Report', *Foundations and Trends in Entrepreneurship*, **7** (1–2), 1–149.

Saxenian, A. (1994), *Regional Advantage Culture and Competition in Silicon Valley and Route 128* (Cambridge, MA: Harvard University Press).

Valentine, Don, et al. (2002), Computer History Museum 'Pioneers Lecture' (September), http://archive.computerhistory.org.

Zhang, J. (2003), 'High-Tech Start-Ups and Industry Dynamics in Silicon Valley', Public Policy Institute of California.

Zhang, J. (2009), 'Why do Some US Universities Generate More Venture-Backed Academic Entrepreneurs than Others?', *Venture Capital*, **11** (2), 133–62.

Conclusion

13. Academic spin-offs and technology transfer in Europe – concluding insights and outlook

Sven H. De Cleyn and Gunter Festel

13.1 INTRODUCTION

Innovation is especially important for regions with limited natural resources like Europe. But the innovation capabilities of European countries differ widely, with significant potential seen in larger countries like France, Germany, Italy and Spain, while others lag behind. Numerous hurdles hamper the commercialization of innovative scientific knowledge and this impedes economic growth. A technology transfer gap exists and must be bridged in order to successfully translate academic research and development (R&D) results into market applications.

Chapter 1 of this book describes how universities and research institutions have established technology transfer capabilities in the early twenty-first century, including technology transfer offices (TTOs). The emergence of university technology transfer activities and the increased exploitation of academic R&D followed the adoption of the Bayh–Dole Act of 1980 in the US and similar legislative initiatives in most European countries. The aim was to commercialize inventions arising from government-funded R&D. As a consequence, as well as education and R&D, the role of universities broadened to include the commercialization of R&D outcomes through licensing and spin-offs.

Meanwhile, universities and research institutions have recognized the importance of entrepreneurship and have steadily increased their activities in this regard. This has been accompanied by a shift in government policy towards encouraging universities and research institutions to commercialize their R&D results, both at national level as well as in R&D projects funded by the European Commission. The ultimate goal is to support economic and social development (job creation, structural change and regional/national development) through more entrepreneurship grounded in academic R&D results. The chapters in this book confirm that policymakers, and especially universities and research institutions in Europe, have learned significantly in recent decades about methodological approaches to support academic spin-offs: from business idea generation over funding to the build-up of organizational structures (and other mechanisms that enable the transfer of knowledge and technologies into market applications). As a result, there has been a substantial increase in the number of academic spin-offs.

TTOs in Europe now recognize that academic spin-offs are a viable method of knowledge transfer and an important mechanism to close the technology transfer gap. They therefore actively assist scientists in their entrepreneurial efforts by establishing entrepreneurship centers and incubators. This book contains several examples, such as initiatives at TU Berlin (Germany) and Linköping University (Sweden), and at research centers like the Fraunhofer Society and the Max-Planck-Society (Germany).

13.2 LESSONS LEARNED

Each of the 12 chapters offers a unique perspective on technology transfer activities and academic spin-offs. The degree of intimacy with which the authors know their subject yields significant insights. Four concrete insights have emerged.

Insight 1 – Funding academic spin-offs remains a challenge
Academic entrepreneurs still face significant challenges in developing their technology and generating revenue early on, not least because of a lack of commercial experience. Insufficient capital and a dearth of industry knowledge present serious challenges for many spin-offs. The resources required in the early stages of high-tech spin-offs are particularly onerous due to the high cost of R&D and product development. TTOs can only address this to a limited extent and external investors are necessary. Publicly subsidized seed investment mechanisms such as the Hightech Gründerfonds can help here.

Insight 2 – New team concepts may be needed, given the relatively inflexible academic career path in Europe
Another challenge is that before and even after the founding of a start-up, the scientist remains absorbed in his/her daily R&D duties and challenges. They may also have a biased view on how the R&D output could be used. In most cases, the entrepreneurial academic is expected to remain a full-time employee of the university/research institution and in a role that

involves more advisory functions, especially technical ones, than hands-on, day-to-day management. Since academic researchers have neither the knowledge nor the experience to commercialize their R&D results, additional operational support is necessary. In such cases, the founding angels approach can significantly help to enable successful spin-off activities, as illustrated in Chapter 9 by Gunter Festel.

Insight 3 – Experienced entrepreneurs and students may be a part of the solution
Given these challenges, many (European) universities and research institutions are experimenting with new approaches to nurturing an entrepreneurial culture and improving the impact of their technology transfer activities. For example, Aarhus University (Denmark) is successfully stimulating student-led spin-offs. The founding angels concept used in Switzerland and Germany is another useful instrument in the technology transfer toolbox, enabling spin-offs to combine the best of both worlds, i.e. academics retaining their positions at the university and business people leading the spin-offs, backed by the scientists. Similarly, the flipped knowledge transfer model of iMinds (Belgium) demonstrates how start-ups led by non-academics can be an enabler of transferring knowledge and technologies into market applications more quickly and efficiently.

Insight 4 – Coaching entrepreneurial academics appears critical
Research and entrepreneurship require different skill sets. Several chapters, including those on Linköping University's Entrepreneurship and New Business Program and on Fraunhofer's Venture Lab initiative, highlight the importance of supporting researchers in their entrepreneurship journey from a human capital perspective. The ambition should clearly not be to turn every researcher into an entrepreneur. However, for those who have the ambition to pursue their own spin-off, the university or research institution can provide programs to help them develop entrepreneurial skills.

On the other hand, researchers may not always be the best people to lead spin-offs. In such cases, this book provides some examples of other solutions, including student-led spin-offs and founding angels.

13.3 CONCLUSIONS

All the approaches and developments described in this book offer reassurance that Europe will increase its innovativeness and consequently its competitiveness in a globalized world. Nevertheless, there are still some fundamental hurdles to overcome (most of which fall outside the scope of this book). They include paralyzing bureaucratic structures, regulatory and administrative complexity, high taxation, and in particular, risk averseness among institutions, banks and other investors as well as potential founders. Therefore, Europe has still not been successful in establishing a Europe-wide entrepreneurial ecosystem – despite the high priority given to innovation on the European political agenda and the large number of support initiatives already in existence.

The further strengthening of entrepreneurial activities in Europe must be based on a holistic approach. This should start with entrepreneurship courses in school and universities, especially for students studying natural sciences and engineering. Spin-offs need good general conditions to flourish, including acceptance by society. Public relations activities should emphasize the value of entrepreneurship (and self-awareness/autonomy in general) for society. The bureaucratic and fiscal conditions should certainly be improved. This would radically increase the impact of different initiatives and programs already supporting entrepreneurship and foster economic growth.

Despite these important hurdles, this book has attempted to show how numerous universities and research institutions across the European continent are successfully developing programs and tools to accelerate the transfer of newly developed knowledge and technologies into market applications. Regardless of the unique context or culture in which they operate, these examples can act as a source of inspiration for other universities and research institutions in Europe and beyond. Just as the economic future of Europe depends on continuous innovation, so does the technology transfer toolbox.

Index